Praise for
TECH TITANS OF CHINA

"This landmark book explores and explains how China is rushing to define the future of commercial technology."
—James McGregor, author, *One Billion Customers: Lessons From the Front Lines of Doing Business in China.*

"*Tech Titans of China* is the most complete and updated chronicle of China's meteoric rise from imitator to innovator in the past decade."
—Kai-Fu Lee, chairman and CEO, Sinovation Ventures and author of bestseller *AI Superpowers*

"Those who do not learn from their competitors are doomed. We need this book."
—Ken Wilcox, Chairman of the Board, Asia Society Northern California, and Chairman Emeritus, Silicon Valley Bank

"*Tech Titans of China* is a fascinating read, and will quickly become the go-to resource to know who the key players are and how China's tech sector is giving the US a run for its money."
—Dorinda Elliott, SVP, China Institute; former Beijing bureau chief, *Newsweek*

"Very few people know more about the rise of China as a competitor to the US than Rebecca Fannin. *Tech Titans of China* clearly lays out the battle for world supremacy."
—Harry Edelson, Chairman, China Investment Group

"Rebecca Fannin has done us all a great service, to de-mystify the Chinese tech scene."
—Craig Allen, Presidcil

"A great informative read on China's technological prowess without the cost of hiring your own consultant."

—Ronald M. Schramm, PhD; Author, *The Chinese Macroeconomy and Financial System;* Visiting Associate Professor, Columbia University School of International and Public Affairs and CEIBS, Shanghai

"This energetic and entertaining book guides us through the Chinese tech jungle. With the US and China facing off as tech rivals, this book is essential reading to understand what is at stake."

—John L. Holden, Senior Director, McLarty Associates; past president, National Committee on US-China Relations

"The must-read book for anyone who cares about where tech is about to go over the next decade and introduces to us how the Chinese ecosystem works and the players in it."

—Robert Scoble, Futurist

"Rebecca's *Tech Titans of China* gives us all an insiders' ring side seat into this dynamic and growing power."

—Brian Cohen, Chairman Emeritus, New York Angels

"Rebecca Fannin's insightful and timely analysis of the rise of Chinese tech and innovation is must reading for every entrepreneur and investor. If you want to understand what is at stake in the looming trade war between the US and China and how it will definitely transform your business, then get this book now."

—Jon Medved, CEO, OurCrowd

"*Tech Titans of China* is a must read to understand the entrepreneurial dynamics and business model innovations of China's leading tech companies and how they will transform the world as they go global."

—Poh Kam WONG, Professor, NUS Business School and Senior Director, NUS Enterprise

"More often than not, Rebecca Fannin has the first line on players in China's 'new economy' for those of us keeping score at home."

—Tim Ferguson, Former Editor, *Forbes Asia*

"Global innovation expert and China guru Rebecca Fannin has penned a brilliant, information-packed, cogent and engaging volume that is a must-read."

—Jerry Haar and Ricardo Ernst, professors respectively
at Florida International University and Georgetown, and
co-authors of *Innovation in Emerging Markets*

"*Tech Titans of China* gives us a very clear and concrete China innovation 101, and lays out how China tech giants are creating their own technology universe."

—Edith Yeung, Creator of China Internet Report,
Partner at Proof of Capital and 500 Startups

"If you care about the future of innovation, then this is an absolute must read book."

—Mike Grandinetti, Faculty Chairman:
Rutgers "Leading Disruptive Innovation" Executive Program;
Faculty Member: MIT Enterprise Forum Startup Founders
Program; Global Professor: Entrepreneurship, Innovation &
Marketing, Hult International Business School

"Whatever the debate may be about Chinese government policies, no one should believe that China's private companies aren't innovative. Rececca Fannin's new book couldn't be more timely."

—Sean Randolph, Senior Director,
Bay Area Council Economic Institution

"An eye-opening book."

—Vivek Wadwha, Professor and Distinguished Fellow,
Carnegie Mellon University Engineering

To Mr. + Mrs. Hughes,

TECH TITANS OF CHINA

*How China's Tech Sector Is Challenging the
World by Innovating Faster, Working
Harder, & Going Global*

Rebecca A. Fannin

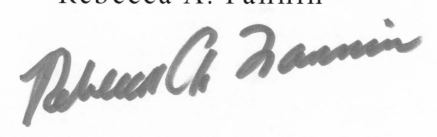

NICHOLAS BREALEY
PUBLISHING

BOSTON • LONDON

First published in 2019 by Nicholas Brealey Publishing
An imprint of John Murray Press

An Hachette UK company

24 23 22 21 20 19   1 2 3 4 5 6 7 8 9 10

A CIP catalogue record for this title is available from the British Library.

Library of Congress Control Number: 2019940692

ISBN 978-1-52937-449-0
US eBook ISBN 978-1-52937-451-3
UK eBook ISBN 978-1-52937-450-6

Printed and bound in the United States of America.

John Murray Press policy is to use papers that are natural, renewable, and recyclable products and made from wood grown in sustainable forests. The logging and manufacturing processes are expected to conform to the environmental regulations of the country of origin.

John Murray Press Ltd
Carmelite House
50 Victoria Embankment
London EC4Y 0DZ
Tel: 020 3122 6000

Nicholas Brealey Publishing
Hachette Book Group
53 State Street
Boston, MA 02109, USA
Tel: (617) 263 1834

www.nbuspublishing.com

To my family, both near and far
and
To the Silicon Dragon global community

Contents

Chapter 3

A next group of up-and-comers is right behind China's BAT and leading the future for smartphones that rival Apple, internet-connected smart homes, and superapps for speedy on-demand takeout lunches, plus 15-second video thrills and AI-fed news.

Chapter 4

It's the rare American internet company that has succeeded behind the Great Wall of China, but Starbucks, Airbnb, WeWork, and LinkedIn keep trying harder with digitally savvy strategies borrowed from China and localized teams.

PART TWO

A core group of US-China leaders set high standards as a Silicon Dragon-style VC market rises to challenge Silicon Valley. A look at who's scoring in China venture investing, and why.

Chapter 5

China's red-hot venture capital market has risen to nearly match the US level and no longer looks to California's Sand Hill Road for cues. Top VC firms in China are funding game-changing innovations at unicorn valuations and achieving high marks for performance.

PART THREE

Chapter 9

The Detroit of Electric Vehicles: China

Buy an electric car in China and you'll receive a free license, a $10,000 subsidy, and access to charging stations. China is the world's leading market for all-electric vehicles, led by Tesla challenger NIO and Alibaba-backed Xpeng Motors.

Chapter 10

The Age of Drones and Robots

China has turned big-time to drones and robots for handling lots of tasks humans can't or don't want to do. Chinese drone startup DJI is the world leader, EHang has a passenger-carrying drone, and robotic vacuum cleaners and window washer startups are getting accelerated at HAX in Shenzhen.

Afterword

What China's great leap forward in the global tech economy means for the United States and its future leadership.

Predicting China's Tech Future

Superpowers United States and China are competing for global dominance of world-changing technologies. It's a pivotal moment. No country stays in power forever.

INTRODUCTION

*China's Silicon Valley has evolved over the past two decades to be
a potentially dominant worldwide tech leader in the near future.
From copiers to originators, Chinese tech titans are showing the
way forward with leading-edge advances that rival the West.*

"Do you use WeChat?" I get this question all the time. Yes! I use
WeChat a lot, in China and sometimes in Silicon Valley. It's the
easiest way to stay in touch with my Silicon Dragon community of
US-China entrepreneurs and venture investors. With China-created
superapp WeChat, you can text or chat in groups or one-on-one;
transfer cash to peers; pay bills; get a loan within seconds; buy movie
tickets; find nearby friends; order groceries; shop for fashions; and
post videos, news, emoticons, and photos. No need for a business
card. You just exchange a barcode-like QR (quick response) code
on your smartphone and, presto, you're connected. WeChat is super
innovative—it combines the functions of Facebook, Twitter, Skype,
WhatsApp, Instagram, and Amazon.

WeChat has more than 1 billion users worldwide, and it's hard
to beat it for work or for play. A San Francisco venture capitalist
completed a term sheet for an investment deal in Beijing entirely
on WeChat. Fans surrounding a Bay Area venture investor speaking
at a Shenzhen conference connected instantaneously with him by

scanning his WeChat QR code from their smartphones. Even beggars in China's major cities carry smartphones with QR codes to receive donations. Cash and email are things of the past in China.

WeChat is just one of many Chinese innovations that is revolutionizing the future with advances that are still rare in the West. China's e-commerce startup Pinduoduo makes online shopping on your mobile for bargains truly social and fun. China's 15-second video streaming app TikTok amuses tweens and can make online performers into rich celebrities—it's what comes after YouTube and Instagram. The world's most valuable artificial intelligence startup SenseTime uses facial recognition on city streets for public security checks. China's electric carmaker NIO stands a chance in its home market of beating Tesla.

China is creating a tech universe that is a counterweight to the long dominance of the United States. In many sectors—mobile payments, e-commerce, electric vehicles, and livestreaming—the Chinese are far ahead. In other deep tech sectors, such as the semiconductors that power electronics and smartphones, American chip makers beat China's national champion.

China has a history of copying Western technology ideas. No more. China beat the United States in landing the first spacecraft on the moon's far side. A Chinese scientist claims his research led to the world's first gene-edited babies. The entire bus fleet in Shanghai and China's southern tech hub Shenzhen is electric. A robot dentist in China was the first to insert 3D tooth implants in a patient. Facial recognition systems catch jaywalkers on LED screens at intersections and scold them on WeChat. Sensors on lampposts collect and send pollution data to the government, which issues advisories to stay indoors when pollution is too severe. China engineered the world's longest sea bridge (over 34 miles) connecting Hong Kong, Macau, and the mainland Chinese city Zhuhai, with a price tag of $20 billion.

China's tech inventors of the future are on fire. For American business and political leaders, it is no longer a good bet to ignore China as the nation advances technologically and seeks to reclaim its once world-leading economic position of centuries ago. China entrepreneurship is hot and showing no signs of cooling down. While trade and technology leadership issues heighten, China innovation accelerates. China's game-changers are already on par with and, in some cases, ahead of the United States. Failure to acknowledge this shift will mean losing advantage to a fast-moving, cutthroat competitor in a high-velocity digital market.

Predictions that China could win a global tech race were once considered laughable. But in the span of little more than a decade since I wrote *Silicon Dragon*,[1] which was the first chronicle of China's emerging Silicon Valley, the world's second-largest economy and its expanding tech empire can no longer be underestimated.

Today, young people in China looking for role models think of Robin Li, Jack Ma, and Pony Ma (founders of Baidu, Alibaba, and Tencent, respectively) more than they do Amazon's Jeff Bezos, Facebook's Mark Zuckerberg, or Steve Jobs of Apple.

High-tech China is inventing the next new thing at a rapid clip in frontier technologies: artificial intelligence, biotech, green energy, robotics, and superfast mobile communications. China also is angling to get ahead in fifth-generation wireless standards, which is being compared in impact to the invention of the Gutenberg printing press.[2]

Large sweeps of the Chinese economy—transportation, commerce, finance, health care, entertainment, and communications—are being reimagined and reshaped by China's assertive effort to forge ahead by leveraging technology. Chinese consumers today are demanding the latest gadgets, and more of them can come up with the dough to pay for the iPhone X than ever before.

China's high-tech giants have dominated in their hyper-competitive homeland for several years. Now that they have gained insights into how Silicon Valley does it, they ambitiously seek to use their knowhow, capital, and scale to become fire-breathing dragons globally. They are stepping onto the world stage and getting recognized by Wall Street, Main Street, Capitol Hill, academia, and the media. The Chinese have arrived and are succeeding at:

- buying into US cutting-edge startups, coinvesting with Sand Hill Road venture capital firms, and spanning out to Southeast Asia and Israel to counter American domination
- fast-tracking and popularizing innovative business models in China that the West is copying: virtual gifts, social commerce, AI-powered news and video apps, and one-stop superapps
- building giant consumer and enterprise ecosystems of mobile payments, online shopping, deliveries, games, and videos that new entrants can't penetrate
- owning and using technologies for smart cities, smart homes, smart workplaces, and smart cars
- inventing the future for mass commercialization of electric cars and self-driving, errand-running humanoid robots, and combined AI and big data to improve cancer diagnoses and care

China has shed its image as the world's low-cost producer and flagrant copier of Western internet and mobile brands to become a breeding ground in today's tech-centric world for disruptive breakthroughs not seen to such an extent since the Industrial Revolution of the eighteenth and nineteenth centuries. China's scale, innovations, speed of execution, and determination to rejuvenate the country and reclaim its glory for the riches, power, and exotic goods that so amazed Marco Polo in his travels to China during the Middle Ages are unmatched. China's innate inventive talent gave us silks, gunpowder, paper, the compass, movable type, and the abacus—and tomorrow, it will unlock more discoveries. Its entrepreneurial clusters in Beijing, Shanghai, Shenzhen, and Hangzhou are challenging

America's Silicon Valley, Silicon Beach, and Silicon Alley for inventiveness, speed, and commercialization of the future for work, play, lifestyle, and connections.

For the first time, China tech innovations are getting ahead of Silicon Valley, and at a fast tempo. Some examples:

- Facebook imitates private group messaging features from Tencent's ubiquitous social messaging app WeChat.
- Chinese drone maker DJI, located in the southern city of Shenzhen, is best in class and the world market leader.
- Alibaba's "New Retail" brings digital technologies and robots into the marketplace to make shopping more convenient and efficient—ahead of Amazon and Walmart.
- Huawei high-end smartphones have been compared favorably with the iPhone X for advanced features and cost and have been popular worldwide—though the China-made phones are effectively banned in the United States due to security concerns and face blocks to Android updates.
- Apple has copied the business model of Chinese smartphone maker Xiaomi by loading up iPhones with subscription content.
- Personal drones to fly passengers short distances as if the Jetsons weren't just a fantasy, thanks to Chinese startup EHang.
- Battery-swapping stations for electric vehicles in China quickly extend driving ranges, without the driver lifting a finger, but haven't gone mainstream in the United States yet.
- China's WeChat Pay and Alipay mobile payments beat Apple Pay and Google Pay for usefulness and adoption.
- On-demand ordering and delivery of takeout orders by super-speedy scooters are replacing restaurant dining and preparing food at home, a growing trend in China that is starting to spill over into the United States.
- WeChat messaging and chats have almost eliminated email and phone calls in China.

In China's fast-moving digital markets, new business ideas and online fads catch on quickly. Mobile apps can fade within a week among a restless, young generation who know what they want and don't want.[3] China's mobile, savvy millennials and Gen Z youngsters mostly in their teens, twenties and thirties, who make up more than half of China's population compared to one-third in the United States, quickly take to all kinds of new techie ways of doing things.[4] For instance:

- It took Chinese messaging app WeChat only three years to reach half the country and now almost everyone in China.
- Giving virtual gifts to online content creators, celebrities, and key opinion leaders happened a couple of years ago in China's booming video streaming market, and the concept is only now catching on in the United States.
- China's dockless bike-sharing business originated in Beijing and Shanghai, and shared bikes popped up more than a year later in Seattle, San Francisco, and Washington, DC.
- A Sesame Credit system scores individuals based on financial status, such as balances held in Alipay accounts, number and influence of online friends, bill payment timeliness, and other data to give creditworthiness scores that can be used for personal credit and even certain social benefits, like premium seating in an airport lounge. China's less restrictive data privacy rules and acceptance of "Big Brother"–like monitoring systems could be adapted outside China (but probably not the United States).
- Alibaba's mobile shopping app lets new-car buyers be identified by facial scans and pick up their vehicle from a five-story, unmanned vending machine. Could this work in Detroit?
- Coffee is ordered by mobile app and then speedily delivered in spill-proof cups for a small fee from battery-powered

mopeds on the fastest routes configured by AI, while bicycle messengers are still the norm in New York City.

- Apps probe data from smartphones, such as battery life and frequency of unanswered calls, and can algorithmically determine an individual's credit-worthiness and automatically issue microloans to previously unbanked people—a practice that US privacy laws would prohibit.
- Supercharging stations for electric vehicles are being built out across China's major cities by the government, while their use remains limited in the United States.
- Short video entertainment apps with low-brow content are popular among farmers and laborers in remote, rural areas of China, while the internet still doesn't penetrate all of Appalachia.

Over the past two decades, I've been closely observing and documenting China's transformation to a switched-on entrepreneurial culture, bigger and faster moving than Silicon Valley. I've spent many days at bustling Chinese incubators, accelerators, conferences, workshops, and networking events—including Silicon Dragon's own events in Beijing, Shanghai, Hong Kong, and Taipei—that foster the exchange of ideas and help startups get up and running. I know and have interviewed China's forward-thinking, risk-taking venture capital leaders and resourceful entrepreneurs who are jump-starting young businesses that have gone on to be big winners. I've seen China's venture and tech ecosystem spread out nationally and strengthen from its original base in Zhongguancun Software Park next to Beijing's well-regarded, tech-oriented Tsinghua University. This story of China's tech economy future continues to be fascinating to experience and cover, particularly from a Western, journalistic perspective.

"China is going to eat Silicon Valley's lunch. It does not make me happy to say that. Pick any single area you want, China is on the right side of the equation. You can argue some of it is because of government protection but at the end of the day, entrepreneurs work harder in China than in the United States; there is equal amount of money; four times the population; and four times the domestic consumer market as that market becomes middle class." That's the strong view of Gary Rieschel, founding managing partner of Shanghai-based Qiming Venture Partners, a leading venture firm actively investing in both the United States and China. He goes on to point out China's progress in biotech is on par with the United States and that China will inevitably dominate the electric vehicle and autonomous driving markets. "If things don't change, the scale of the market, the number of graduates, every single aspect of the infrastructure that China has put together, in 10 years we will be talking about, 'how do we save the entrepreneurial spirit in Silicon Valley.'"

> *"China is going to eat Silicon Valley's lunch. It does not make me happy to say that."*
>
> **—Gary Rieschel**
> Founding managing partner, Qiming Venture Partners

Rieschel knows full well the key factors driving China's push to win the global tech race, from working and living in Shanghai for eight years on the front lines of China's techno-charged environment. China's workaholic, determined entrepreneurs are quick to get new technologies commercialized. Chinese people voraciously embrace the latest apps, games, payment services, social media, and online shopping. Venture capitalists fund cutting-edge startups in artificial intelligence, self-driving cars, electric vehicle batteries, biotech, robotics, drones, and augmented and virtual reality. China's huge digital markets—the world's largest for the internet, smartphones, e-commerce, and mobile payments—are spurring advancements that go mainstream quickly. Not least of all, the Chinese government's protectionism and concerted nationalistic policies propel China to become a world-leading innovative country. Don't blink or you will miss this growing challenge to America's technological power.

Go to China and see for yourself. You can't help but be impressed by China's high-speed national railways and smoothly paved, multilane highways that connect huge cities, long-span bridges, new airports, tall skyscrapers, glistening shopping malls, palace-like corporate parks, and professionally designed coworking spaces. China's clean and contemporary central business districts make hubs in New York City, San Francisco, and Los Angeles look dirty and dated. I can recall when Shanghai and Beijing construction cranes dotted the cityscape; when oxen-drawn carts came into Beijing from rural areas; and when China's internet startups were housed in dingy walk-up buildings. Now those startups are located in ultramodern corporate campuses. I remember when there were only a few Westernized hotels in Beijing, and now Hilton, Hyatt, and Marriott have many luxury properties. It wasn't long ago that the Friendship Store was the only place to buy peanut butter. Today, McDonald's and Kentucky Fried Chicken are everywhere. I see Nike shoes and Coach handbags being sold in upscale shopping malls. I remember when bikes were the main way of getting around and well recall when China's new car owners made traffic jams so terrible it took two hours to get to an appointment across town. Now shared bikes, scooters, and mopeds clog the streets. AI-powered surveillance cameras positioned at intersections monitor traffic flow, collect data, and track people. Architectural wonders stand out, like the CCTV pants-shaped building in Beijing, the bottle opener–shaped Shanghai World Financial Towers, and the aptly nicknamed Bird's Nest stadium, built for the Beijing Olympics. How could I forget that few in Silicon Valley wanted to hear about China's catch-up to the West. Now, it's talked about a lot, in a worrying way.

Over the past few decades, the world's second-largest economy has pivoted its central long-term economic strategy from manufacturing and exports to consumer goods, and now a techno-nationalism. While China is leading in many of tomorrow's advances, it trails in fundamental technologies, such as semiconductors, where the country relies on foreign high-tech know-how.[5] China, however, is on a mission to close that gap.

Leapfrogging to Copied from China

In less than two decades, China tech innovation has evolved and gone through three phases of development: from copy to China, to invented in China, and now today, the biggest trend to watch is copied from China, meaning US companies duplicate Chinese innovations.

China's first-generation internet entrepreneurs unabashedly created copies of successful American startups Yahoo!, Amazon, Facebook, Google, and eBay. Intellectual property protections were scant. Now Chinese technocrats are breaking boundaries with their own disruptive innovations, taking them overseas and getting copied by Westerners. Chinese internet companies are no longer dismissed as mere copycats of Google, Facebook, YouTube, and Amazon— as they were when my *Silicon Dragon* book first was published in 2008.

Now, made-in-China business models built for a mobile-first generation are advanced and widely used. There are multifunctional superapps, mobile wallets, mobile shopping in groups, mobile videos and streaming, mobile books, and mobile news apps with no editor. This world-on-a-screen is extended by social sharing features baked into these apps.

No other country in the world can match China's startup zeal. Founders in Beijing, Shanghai, Hangzhou, Shenzhen, and China's tier-two cities are tireless, persistent, and driven to succeed. No fear of failure here; it's fear of missing out. Entrepreneurs and venture investors schooled in the West at elite universities such as Stanford,

Harvard, Princeton, and Yale, and PhDs trained at top engineering schools like MIT, Caltech, UC Berkeley, and Carnegie Mellon keep returning to China to make their startup mark. Managers with international experience are recruited from well-known US tech companies to drive global expansion of Chinese companies from Beijing—at least until their lungs protest from the polluted environment. Startup teams in China routinely work 12 hours per day, six days a week, or "996," as it's commonly referred to in US-China tech circles. It's a reminder of Silicon Valley all-nighters during the late 1990s dotcom boom when China's entrepreneurial boom was only percolating. "China and the US are at different points of economic development and motivation. China's entrepreneurial culture does make Silicon Valley look sleepy," says Hans Tung, managing partner at leading venture investment firm GGV Capital in Menlo Park. Mike Moritz, partner at top-tier Sequoia Capital, can't help

> *"China's entrepreneurial culture does make Silicon Valley look sleepy."*
>
> **—Hans Tung**
> Managing partner, GGV Capital

but agree. He points out that Chinese entrepreneurs who routinely work 80 hours per week are making their Silicon Valley peers look "lazy and entitled."

When traveling to China, as I've done more than 100 times for work, I'll often have breakfast or lunch meetings on weekends in Beijing or Shanghai. That rarely happens in Silicon Valley, where skiing, golfing, sailing, and walking over the Golden Gate Bridge fill up the weekends.

Necessity being the mother of invention, what's propelling China today at full speed ahead toward global tech leadership?

- China's Soviet-style thirteenth five-year plan for 2016–2020 to boost economic development by accelerating indigenous innovation, mass entrepreneurship, research and development, and patents to burst open as an innovation nation by 2020 and a technological and scientific world powerhouse by 2050—timed to the 100th anniversary of the Communist Party of China.

- A state-led blueprint, "Made in China 2025," to close the gap in technology leadership by building national firms into globally competitive tech champions and gaining technological leadership in emerging sectors including robotics, new-energy vehicles, biotech, power equipment, aerospace, and next-generation information technology—all to achieve supremacy.[6]

- The nation's "Internet Plus" plan to build up China's companies as world-class competitors in mobile internet, big data, cloud computing, and Internet of Things.[7] The proposal's focus on optimizing health care, manufacturing, and finance by leveraging internet connectivity and big data.[8]

- Chinese president Xi Jinping's Belt and Road initiative to build a twenty-first-century Silk Road land and maritime trade corridor that could outdo America's postwar reconstruction Marshall Plan to foster economic integration with neighboring countries, boost demand for Chinese products, and develop China's poorer western provinces.

- A state-led $15 billion China New Era Technology Fund to invest in startups and cutting-edge technologies[9]—and to acquire expertise from abroad when indigenous development is not possible.[10]

- China's venture capital market that nearly equals Sand Hill Road in size and impact, and measures up with mega-financings, and growing share of unicorn-financed startups globally.[11,12] Chinese digital content app ByteDance ranks first as the world's most valuable privately held startup.[13]

- Tencent's and Alibaba's rank in the top ten publicly traded companies for market valuation, in the same league as

Microsoft, Apple, Amazon, and Facebook.[14] A decade ago, not a single China internet company made it into this tier.

- China's increasing number of tech companies going public on major stock exchanges. In New York, 31 Chinese companies raised $8.5 billion in 2018, with highly innovative electric carmaker NIO and social commerce upstart Pinduoduo out in front and several more China IPOs have followed in 2019.[15]

- China's push to put their stamp on technology startups overseas. From 2010 to 2018, Chinese deal makers made 1,315 tech investments globally, investing $99.8 billion, with a considerable chunk in the United States.[16]

- China's gains in national R&D spending to $409 billion, catching up to the United States at $497 billion,[17] and predicted by the National Science Board to soon surpass the United States.[18,19]

- Huawei's fifth-placed spot worldwide in R&D investment, ahead of Intel and Apple.[20]

- China's climb in patent applications—from seventh place globally a decade ago to second today,[21] at 21 percent of the total worldwide, not far behind the United States, the longtime leader at 22 percent. China is predicted to surpass the United States for patent filings within two years, if current trends continue. Huawei's lead, by far, as the world's top filer for new patents.[22] China's rank as second worldwide for patents in use—29 percent of a total 13.7 million patents compared to 40 percent for the United States.[23]

- China's 4.7 million graduates in science, technology, engineering, and math, outnumbering 568,000 for the United States, and predictions that China's STEM graduates will rise 300 percent by 2030.[24]

- China's new lead over the United States in academic scientific research papers to 18.6 percent of the world's science and engineering publications—426,000 papers from China as opposed to the United States at 409,000.[25]

- China's leading share of the top 500 speediest supercomputers globally—202 compared to 143 in the United States—an important part of China's grand plan of technical prowess.[26]
- China's world-leading number of internet users (829 million) and smartphone users (783 million), several times larger than the United States' at 293 million and 252 million, respectively.[27]
- Huge untapped potential since internet penetration in China is 58 percent compared with 89 percent in the United States.[28]

While China's sheer size and fast economic development means that it leads in many indicators, it's undeniable: China is on a tech upgrade that will challenge the West for leadership of the global economy for the coming decades just as America dominated the industrial and information revolution in the past century. A shake-out will occur if Silicon Valley doesn't recognize and respond to these leading signs of a massive power shift. While the United States is king of the tech hill, other Silicon Valleys have sprung up in Tel Aviv, London, Bangalore, and elsewhere—but most powerfully in China.

As China doubles down on its goal to achieve supremacy with national high-tech champions that can compete globally, separate spheres of power are forming: China in Asia and the United States in the West. China's tech giants are getting ahead quickly in the fast-growing Asian region—proximity and consumer familiarity with Chinese brands and digitally advanced features helps. The tech universe of the future will be regional innovation powerhouses moving in parallel with neither dominating worldwide.

One thing US tech leaders can't do much about is the uneven playing field—American companies can't get beyond the Great

Firewall of China's internet censorship and reach Chinese consumers directly. Facebook's Mark Zuckerberg meets with Chinese leaders, jogs in front of the Forbidden City, and is learning Mandarin, but Facebook remains off-limits in China. Blocks remain on Google, Twitter, and Instagram too—though Google has been making plans to reenter China with its controversial Project Dragon, a censored search app for China, in the face of White House pressure to scrap it. The overall effect is that Chinese internet brands own the market in this huge tech arena. I can't tell you how many times I've grown frustrated with attempting to search for information in China and can't access Google.

Few American tech companies have succeeded in China—the casualties include eBay, Groupon, Google, Facebook, and Pinterest. But Tesla, WeWork, Airbnb, LinkedIn, Starbucks, and others continue to push into the Middle Kingdom, with mixed success.

IP Theft and Counterfeiting

What could go wrong and derail China's dream to ascend in tech? Plenty. Trade wars and tech power battles. Friction over American companies being forced to play by the Chinese government's rules, turn over key technologies, and compete with homegrown companies that are nationally subsidized. President Donald Trump's deepening push to get tough on China. Increased US tariffs on Chinese goods to deflate a US-China trade imbalance. Stepped-up US export controls of advanced technologies to China. Stricter reviews and blocks by Washington DC, on foreign investment in high-tech American companies. Crackdowns on cybertheft and disregard for intellectual property, such as the recent US criminal charges against Huawei. Increased visa restrictions by the United States on Chinese national graduate students in science-related fields. Blocks on China's heavy reliance on American-designed, higher-end technologies from suppliers such as Google, Qualcomm, Marvell, and

Intel—issues that came to the fore recently when a Chinese man-ufacturer was accused of swiping American-designed chips from Micron Technology to build a new Chinese plant[29] and when Chi-nese telecom giants ZTE and Huawei were prevented from buying American supplies because of national security issues. These inci-dents have firmed up China's resolve to cut reliance on US tech smarts to fill gaps and to grow its own core technologies, but it will take years and won't be so easily done.

A host of issues could put the brakes on what fuels Chinese startups: curbs on China venture capital in US tech startups and a higher bar for Chinese companies to go public in the United States and use that capital to scale up in China.

Geopolitical issues loom large. Nationalization or breakups of China's tech titans. Military conflicts over pending trouble spots in the South China Sea and China's claim to Taiwan. Poorly imple-mented state-backed reforms on a local level. Growing criticism over China-styled colonialism, such as using loans to gain control over strategic locations, notably the Sri Lanka port and surround-ing land. It's conceivable that China could roll back the capitalis-tic reforms ushered in by Chinese leader Deng Xiaoping in the late 1970s to return to Chairman Mao's drab communism of several decades ago.

National security and technological leadership frictions will no doubt heighten as China marches forward. Technologies we use every day—Siri, touchscreen, GPS, the internet, and the iPhone—came out of the US Department of Defense and government-funded scientists for military purposes.

"Anyone who is trying to understand China today without understanding what's happening from the innovation perspective or what's happening to China's entrepreneurs, is not thoughtful," says tech investor Gary Rieschel, a thought leader in the world of venture and entrepreneurship. "That's a huge blind spot for a great number of policy specialists in United States working on China."

Merit Janow, dean at Columbia University's School of International and Public Affairs and a former deputy trade representative to China and Japan, gives her expert perspective: "Trade policies won't fix the fact that China is investing heavily in tech R&D and education, while the United States is cutting. China innovation is gaining with large clusters of tech hubs and government support."[30]

> "Trade policies won't fix the fact that China is investing heavily in tech R&D and education, while the US is cutting."
>
> **—Merit Janow**
> Dean, Columbia University's School of International and Public Affairs

China's race to be the global leader in upcoming fifth-generation (5G) wireless mobile communications is one of the key battlegrounds. This new breakthrough will turbocharge connections and hyperconnect data from smartphones, laptops, refrigerators, dog collars, medical devices, and vehicles—and transform how homes, cities, hospitals, vehicles, and factories operate. Racing to get a first-mover advantage, China has outspent the United States on wireless infrastructure and cell sites by approximately $24 billion since 2015 and plans to invest more than $400 billion in 5G testing and development over the next 10 years.[31] What's standing in China's way is US and foreign ally blocks of China-made core equipment in national networks over security concerns. The ban could lead China to develop stand-alone 5G networks and result in a fragmented market and higher prices for consumers and telecoms.[32,33]

Underpinning China's tech economy gains are a whole host of socioeconomic issues and cultural issues that could deter China from what it considers its rightful place as a world-leading economic power, one of the richest, most prosperous, and most industrious countries in the world until the nineteenth century. Teaching that has prioritized memorization and test taking in China instead of creative thinking. An aging society and declining workforce after decades of the Chinese government's one-child policy, which was eased to two children by 2016 but still has resulted in lagging birthrates. Censorship of information and blocks of American net brands in China that restrict knowledge, hinder creative expression, and promote conformity rather than the out-of-the-box thinking that drives innovation. And to top it off, China has to deal with pollution, income inequality, and Chinese banks' nonperforming loans.

Whether you are a panda lover or dragon slayer, it's hard to ignore Chinese tech ascent into the spotlight. China now represents about one-third of exhibitors at the mega annual Consumer Electronics Show in Las Vegas, where dozens of companies from all over the world display their wares. Most of China's top tech executives speak English fluently, while the reverse is not true—except maybe Zuckerberg, who is studying Chinese.

Technology Leadership, Sector by Sector

China's startups and tech titans alike are distinctly focused on the most progressive ideas in the digital world today and are challenging the United States in several sectors, with China-for-China innovations and business models.

Examples include:

- **Artificial intelligence:** Baidu is emerging as a leader in AI with driver-less-car technology and AI voice-activated smart-home devices. The United States has the technical lead, but China has the ability to innovate faster, given large data sets. It's fifty-fifty over which market—China or the United States—will get the edge.
- **New retail commerce:** Alibaba and JD.com have pioneered cashless and cashier-free stores and are digitalizing China's retailing and logistics with efficiencies in merchandising, pricing, and marketing. They are also transforming

deliveries with white-glove service and super-speedy scooters. Alibaba's futuristic Freshippo grocery stores employ robots and are more advanced and extensive than Amazon Go's limited number of automated convenience stores in the United States.

- **Mobile payments:** China today is a cashless society. China's mobile payments market led by WeChat Pay and Alipay already exceeds US credit and debit card usage.
- **Fintech:** Alibaba affiliate Ant Financial is a one-stop financial services giant that uses big data and machine learning to dominate in money market funds, lending, insurance, mobile payments, wealth management, and blockchain services.
- **Social credit:** China's new, controversial social credit system judges a citizen's trustworthiness through technological surveillance and encourages compliance by giving ratings that can determine access to loans, jobs, schools, and travel.
- **Sharing economy:** China-invented business models for shared bikes, battery chargers, umbrellas, basketballs, and takeout kitchens have been popularized by dozens of startups.
- **Livestreaming:** Video streaming sites from Baidu's Netflix-like iQiyi and digital entertainment innovator YY are booming and creating online celebrities paid in virtual gifts by addicted viewers.
- **Virtual reality:** Arcades across Chinese cities are crowded like US amusement parks and offer VR escapes for about the price of a movie ticket.
- **Electric vehicles:** Makers NIO and Xpeng Motors of full-featured and well-funded Chinese electric cars with self-driving capabilities and built-in entertainment apps are being propped up by government subsidies to promote clean energy driving and push China to be the Detroit of electric vehicles. China has already emerged as the world's

largest maker and seller of electric cars. Dozens of Chinese new-energy vehicle makers have established Silicon Valley centers to upgrade R&D for next-generation all-electric, self-driving vehicles.

- **Social commerce:** Nifty Pinduoduo lets shoppers buy directly on their mobile from merchandisers, get coupons, win prize drawings, and earn discounts when they put together a group purchase with their connections.

Over the past decade, China's great tech leap forward has been supercharged by an abundance of venture capital often sourced from Western pension funds, endowments, family offices, and foundations. Records keep getting set for funding sizes, IPOs, deals, and investment performance (see chapter five, page 127). China's ride-hailing leader, Didi Chuxing, has pulled in $21 billion in funding since starting in 2012, and even took over rival Uber in China in 2016 in a groundbreaking $35 billion transaction.[34] Several Chinese tech unicorns have scored blockbuster IPOs: Chinese smartphone maker Xiaomi nabbed a market capitalization of $54 billion and food services app Meituan attracted $53 billion, while shopping app Pinduoduo took only three years to go from zero to a $24 billion IPO in New York.

More Chinese tech IPOs are coming, and investors eagerly await them. Venture capitalists who are behind the rise of these tech titans have amassed huge wealth—as much as a 30 times' return on $8 million of investment in 2013 into Alibaba. Investors in Alibaba stock have doubled their money since the company went public in 2014. More money will continue to be made as China's tech engine accelerates.

At the forefront of China's tech boom are its homegrown giants Baidu, Alibaba, and Tencent, collectively known as the BAT. They own Chinese search, e-commerce, social media, and gaming and are innovating across broad swaths of the tech economy. Right behind them are another group of Chinese creatives—AI news aggregator Toutiao and video app TikTok, services superapp Meituan Dianping, ride-hailing leader Didi, smartphone maker-plus Xiaomi, and many other up-and-comers in artificial intelligence, electric cars, drones, and more. China's tech giants want power. They are supersizing by merging and acquiring many types of businesses and consolidating them under one roof. In one major recent deal, Meituan acquired Chinese bike-sharing startup Mobike for the large sum of $2.7 billion to add more delivery options to its service.

Acquisition Fever

Going global is the next push of these China-centric tech giants. China's tech titans have invested in and acquired startups and cutting-edge emerging companies throughout leading hubs worldwide, formed Sand Hill Road venture capital units, set up R&D outfits close to engineering talent, and angled into Hollywood moviemaking in a bid for soft power.

In this outward reach, China investment in US tech companies reached $51.4 billion from 2010 through 2018, led by megadeals in America's top trophy startups Uber, Lyft, and Magic Leap.[35] Recent US regulatory hurdles and a Beijing crackdown on high-priced, debt-laden deals have curbed the action. But the innovation engine keeps going, and so does venture capital from Silicon Valley and China funds to fuel it.

In the wake of heightened regulations and uncertainty, Chinese tech deal makers are shifting to smaller, highly strategic transactions in the United States and turning to more welcoming markets internationally. China's tech titans Baidu, Alibaba, and Tencent are pivoting from the United States to "startup nation" Israel and to Southeast Asia, staking out positions ahead of the United States in a region booming with innovation that promises to echo some of China's success.

Power Centers Split: East and West

While America struggles to get it right with making America great again, China's nationalistic, state-led push into the modern age is undergoing its own tensions. Chinese culture has become more assertive, confident, and aggressive. I've noticed stress levels rising. I've witnessed people start to cry when someone beats them to a taxi. I've seen passengers who won't shut down their smartphone when the plane is on an active runway. I've watched construction workers carry an injured construction coworker off of a building site and leave him in the dirt. There have been instances of murders and rapes in China's ride-sharing services. This is not the harmonious China I remember from years back walking along Nanjing Road with hordes of people on a beautiful Mid-Autumn Festival moonlit night.

Sending an Alert

China's shake-up of the status quo sends an alert to Western business leaders and policy makers. China's tech influence and power grows daily. Make America Great Again slogans are up against China's ambitiously planned policies that call for achieving self-sufficiency in tech and becoming a manufacturing superpower that dominates the global market in critical technologies.[36]

Silicon Valley's continued world dominance is at stake. Everyone still goes to Silicon Valley to be immersed in technology and venture capital—and to learn its secrets to success. Not everyone travels to China to discover tech treasures. The United States cannot afford to become complacent. Let's hope that the rise of an emerging power threatening the leading global power will not lead to a Tech Cold War.

part one

HOW CHINA IS WINNING

CHAPTER 1

CHINA'S FACEBOOK, AMAZON, AND GOOGLE: THE BAT'S TECH TEMPLATE

China's tech titans Baidu, Alibaba, and Tencent (the BAT) own search, e-commerce, and social networking in China and are forging ahead into innovating frontier technologies that will reshape financial, retail, transportation, and mobile communication sectors globally.

The BAT's Tech Template

The Alibaba-owned supermarket I visited in Shanghai makes the Whole Foods grocery store in my New York City neighborhood look outdated. There's no cash, cashiers, or checkout lines at Alibaba's futuristic stores. It's all digital. You pay by Alipay mobile app at do-it-yourself kiosks equipped with digital cameras that scan and ID your face. Your groceries orders are bagged and clipped to an overhead conveyor belt that runs across the store to be delivered within 30 minutes by scooters within a three-kilometer zone. Aisles with built-in interactive, digital screens let you see the ingredients of each food and its origin. Wi-fi–connected, e-ink price tags change pricing details automatically. Fresh produce and seafood are flown in daily and can be prepared and eaten at the in-store restaurant, where robots travel along a track to bring your freshly cooked meal directly to your table in a sealed compartment.

Alibaba has opened more than 100 self-operated Freshippo grocery retail stores in major Chinese cities, and it plans more. In the United States, Amazon is playing catch-up with its own e-retailing and deliveries that are not as far-reaching or digitalized yet. Amazon's small, automated Go convenience stores offer much narrower merchandise selections and are only in 10 US metros, while Whole Foods grocery deliveries can take two hours and are limited to select cities, with free service restricted to Amazon Prime members. Neither has the new-fangled tech that I saw in Chinese retail outlets. This is just one example of where the digital future is already here, in China, while the United States lags.

Today, Alibaba stands out as one of China's three tech kings, along with Baidu and Tencent, collectively known as the BAT. Like the FANGs, namely Facebook, Amazon, Netflix, and Google in the United States, in China Baidu owns search, Alibaba leads e-commerce, and Tencent dominates gaming and social networking—and all are well positioned in artificial intelligence (AI). Their success has come from hard work, ambition, talent, capital, and a home team advantage in China's newly entrepreneurial markets. In the scrappy, cutthroat Chinese marketplace, these digital innovators seek to own the next new thing and are creating awesome new features and business models that have captured the Western world's attention. Their next hurdle is breaking through globally.

Alibaba, with a name that conjures up the command "Open sesame," is universally known due to charismatic leader Jack Ma and his e-commerce treasures. Now the massive power and reach of the BAT is increasingly recognized outside of China. All three are at the vanguard of China's turbocharged digital economy, which is set to jumbo-size by five times to $16 trillion in 2035.[1] They are changing how Chinese consumers connect, socialize, shop, pay, eat, travel, invest, get loans, and monitor their health, and are causing America's own innovators to sit up and take notice.

The BAT have the best of both worlds. These three companies have scaled up in their protected home markets with little overseas competition, tapped the Western capital markets with IPOs, and used the money to bulk up with billion-dollar acquisitions and chunks of the most promising, innovative tech companies, in effect paying a large tuition to gain knowledge on what makes Silicon Valley tick. They are innovating in ways that are tuned to China's fast-paced digital culture and slowly filtering into the West. Their sprawling size and take-all competitive thrust has major implications for the US tech industry and for US consumers.

In short, Baidu, Alibaba, and Tencent are upending a long-held and cherished assumption that Silicon Valley companies will dominate the global tech economy in the coming decades. They are leading a high-tech gold rush guided by five key trailblazing strategies:

- snapping up cutting-edge technology startups
- keeping the innovation engines humming day and night
- expanding to high-potential tech hubs around the world
- advancing into artificial intelligence, big data, telemedicine, autonomous driving, facial recognition, mobile payments and lending, and digital entertainment
- constructing moats of many tech sectors in sprawling ecosystems to bulk up and ward off competitive invaders

Li, Ma, and Ma, Oh My

Before taking a closer look at these aggressive initiatives, let's recap how far this trio of giants has come. Over the past decade, China's tech heavyweights have entered the rarefied ranks of the world's largest and most valuable companies. They are sometimes compared with South Korea's chaebol or powerful conglomerates Samsung, LG, and Hyundai.

Isolated from American competition thanks to the Chinese government ban on Google, Facebook, and Twitter, as well as struggles by Amazon, eBay, Yahoo!, MySpace, and others to crack the Chinese market in the face of government restrictions and support of home-grown, internet warriors, China's tech titans have logged remarkable double-digit growth rates for several years—and continue to accelerate.

Enormous progress has been made since 2000, when digital China first arrived, thanks to the foresight and leadership of the BAT CEOs: Robin Li of Baidu, Jack Ma of Alibaba, and Tencent's Ma Huateng, known as Pony Ma (derived from his family name, which means *horse* in Chinese). It's hard to imagine they would have come so far, so quickly—and stayed in power for so long. They are regarded as superheroes in the first generation of entrepreneurs since China's Cultural Revolution of the 1960s and 1970s nearly destroyed the economy, and the later reforms of former leader Deng Xiaoping opened China to a socialist market economy and made it glorious to get rich.

All three founders cranked up their startups soon after the US dotcom bubble burst about two decades ago. They did become very rich—among the richest in the world—initially by copying. Baidu's quiet-spoken search expert Li, who came to the United States for a computer science graduate degree and jobs at Dow Jones and Disney-owned search company Infoseek before returning home, has a $10 billion fortune made in China. He unabashedly copied Google and won the Chinese market. Alibaba's dynamic leader, a former English teacher who acted on his entrepreneurial instincts after discovering the internet on a trip to the United States as a translator, has a net worth of $40 billion. His Taobao shopping site was very intentionally modeled on eBay and beat eBay in China with cutthroat pricing and localized features such as instant messaging.

Tencent's CEO, a press-shy engineer born and educated in southern China, has been described as a scorpion who will lie in wait before attacking, has a $44 billion fortune.[2] His QQ instant messaging service was based on the Israeli invention ICQ, initially acquired by AOL and then bought by Russia's largest internet company, Mail.Ru.

Today, China's titans have left copying behind and are managing broad and deep power bases in tech. With that power comes lots of headaches. Like Mark Zuckerberg, Jeff Bezos, and Larry Page, who confront a tech backlash and constant challenges to their clout, China's leaders face daunting issues that could weaken them: privacy concerns, counterfeit charges, restrictions on their most addictive products, and competitive threats.

Baidu faces a possible reentry of Google to the Middle Kingdom some 10 years after googling didn't knock out China's search leader. Baidu's bid to own the future for AI with self-driving cars and facial recognition for payments is uncertain after Li lost two experts in a row who were leading Baidu's AI charge while rivals chip into the sector: Alibaba in smart-city traffic management, Tencent in medical imaging and diagnostic tools, and startups SenseTime and Face++ with AI-enhanced face-matching technologies for IDs and public security.

Alibaba's lead in e-commerce is challenged too, by social commerce upstart Pinduoduo with an app that combines bargain bin merchandise, prizes, social sharing features, and gaming. Alibaba also must fight a Tencent-funded gang of e-commerce contenders, including giant JD.com.

Tencent has to recapture the entrepreneurial spirit and creative juices that turned WeChat into a multifunctional superapp but missed the short video phenomenon grabbed by newcomer TikTok. Tencent must also reset its main gaming business, which got caught up in Chinese regulations. When I first met and interviewed China's young tech leaders in the early years of the new millennium, they were just getting started in figuring out the magic of Silicon Valley and had little track record. Little did they know it would get so complex, so fast. And neither did I.

How China's Net Brands Stack Up

As China's big three tech companies have forged their paths, they parallel America's FANGs. They are still far overshadowed in size by Google, Amazon, and Facebook but growing superfast. If their growth continues at a rapid clip, China's BAT might someday measure up to American companies. See table 1-1. The Chinese threesome already are among the most valuable publicly traded companies in the world. The market capitalizations of Tencent and Alibaba both hover around $400 billion, in the top 10 with Amazon, Apple, Microsoft, Alphabet, and Facebook, while Baidu trails at about $60 billion.[3]

China's game-changers are innovating, investing, and commercializing quickly, with many next-generation technologies that the West doesn't have yet for the home, office, vehicle, and screen. None of these titans is likely to be dislodged from its position soon, but

Table 1-1

Comparing China's BAT with US Tech Leaders

	2017 Revenues	% increase	2018 Revenues	% increase
Baidu	$13 billion	+20%	$14.9 billion	+28%
Alibaba	$39.9 billion	+58%	$56.2 billion	+51%*
Tencent	$36.4 billion	+56%	$45.6 billion	+32%
Alphabet	$110.9 billion	+23%	$136.8 billion	+22%
Amazon	$177.9 billion	+31%	$232.9 billion	+31%
Facebook	$40.6 billion	+47%	$55 billion	+38%

Sources: annual reports, company news releases

look for shifts. The vortex of China's techno power is increasingly spinning around the aggressive *A* and the *T* of the BAT, Alibaba and Tencent, while Baidu loses ground to its bigger rivals and posted its first quarterly loss in 2019 since a 2005 listing. There's a new *B* on the horizon: the world's highest-valued unicorn, ByteDance, which is taking on Tencent with internationally popular video app TikTok and a new video messaging app.

In the race to own it all, Alibaba and Tencent clash frequently in mobile payments, messaging, and mobile commerce. Alibaba, the master of online services for sellers to become faster, smarter, and more productive, is pushing into Tencent turf in social networking with DingTalk, a messaging app for the workplace. Tencent, the leader of digital content and entertainment, is playing in Alibaba's sandbox by launching WeChat minishops. Baidu is largely staying out of the fray, focused on AI for autonomous driving and smart-home devices for lighting and speakers.

New rivals are ready to pounce on China's BAT companies if they slip up. But it won't be easy to penetrate their league. The clubby network effect in China is just as strong as it is in Silicon Valley. Similar to PayPal's so-called mafia, the BAT trio and their alumni represent 42 percent of China's venture investments, one in five top Chinese startups, and 30 percent of funding in Chinese startups.[4]

Copying from China

China's rough-and-tumble environment toughens up entrepreneurs and forces them to get new ideas into the marketplace fast and first, often ahead of US counterparts, which are today ending up as the copycats. Notably, Facebook has been studying China for clues. Facebook redid its social media site to integrate private messaging, group chats, and payments that resemble innovations that WeChat pioneered years ago.

Facebook tested its own digital gift feature, seeing the quick uptake of a digitalized spin on the traditional Chinese custom of giving red packets stuffed with money for family and friends. Facebook also relaunched short video app Lasso, described in China tech circles as a "100 percent" copy of the Chinese-created TikTok app from ByteDance.

China's massive $9 trillion mobile payments market led by Alibaba and Tencent is light-years ahead of the United States.[5,6] Nearly everyone in the country, about 900 million, uses their smartphone as a mobile wallet, Alipay or WeChat Pay, to scan items and pay instantly without banking and credit card fees. Cash is a thing of the past in China. Cash, checks, money orders, and credit and debit cards are still commonly used in the United States, and Apple Pay and Google Pay are not yet mainstream in the States.[7] Breaking into China now isn't an option for American financial brands, given the dominance of Alibaba and Tencent. Google's app store is blocked, and Apple Pay has such limited traction that Apple CEO Tim Cook agreed to accept Alipay at Apple's 41 retail stores in China. MasterCard and Visa have tried for years to break into China while American Express was recently approved, but it's probably too late. China has bypassed credit cards just as it did the personal computer for mobile phones and retail stores for online commerce.

The B of the BAT

Baidu Goes for AI

Now nearing its twentieth year, Baidu faces growing pains, which is why it's trying so hard to diversify into cool AI technologies from its world-leading Chinese-language internet search engine.

The huge Consumer Electronics Show (CES) in Las Vegas is where global tech companies strut their stuff, seeking to gain a foothold in the $1.6 trillion US tech market. CES has long been a bastion of US tech companies, from Amazon to Google to Microsoft. Yet Chinese companies are starting to take center stage. With its first US press conference at CES in 2018, Chinese search leader Baidu staged a flashy coming-out party. Then Chief Operation Officer Lu Qi showed off the company's AI technologies and boasted that Baidu is "innovating at China speed," or superfast. Hired from Microsoft by CEO and cofounder Robin Li to boost Baidu in AI, the COO was hawking all the latest wares: self-driving technologies to compete with Google and a line of DuerOS voice-powered speakers, lamps, and projectors he not so subtly dubbed the "Alexa of China."

A few months later, at the annual Baidu World forum in Beijing, cofounder Li set an upbeat tone with the program theme "Yes, AI do," as he showed off a series of product introductions and upgrades that proved how far this leading search company is veering from its origins. Dressed in his classic white shirt emblazoned with a Baidu logo, he put a positive spin on Baidu's latest innovations in artificial intelligence—not so easy to be upbeat since he had to step back into the CEO seat after his star hire Lu left in May 2018 to launch and run a China offshoot of US accelerator Y Combinator. And that was only a year after AI superstar Andrew Ng departed Baidu for a new AI mission in Silicon Valley.

As Li spoke, flashy surround-sound videos displayed Baidu's new technologies for autonomous driving, smart-city projects in Beijing and Shanghai, and voice-activated speakers and lights. The crowd at the packed ballroom of the China World Hotel cheered loudly and clapped with each introduction: a pilot launch of 100 self-driving taxis in China's central city Changsha, a partnership with Volvo to develop self-driving electric vehicles for the large China market, and an alliance with China's large auto manufacturer FAW Group to produce autonomous passenger vehicles and start testing them in 2019 in Beijing and the northeast Chinese city Changchun.

While over the top with stage effects, this push by Baidu into artificial intelligence is not so much of a stretch. Like search, the company's original and still biggest revenue generator, its AI technologies rely on computerized algorithms to retrieve stored information and feed sensors for self-driving vehicles and voice-activated lights and speakers.

Now back in the driver's seat at Baidu, the challenge for Li is keeping a big-picture vision while running day-to-day operations and juggling both AI and search businesses.

Next for Baidu is making money from its suite of AI products, branded Baidu Brain; voice-assisted DuerOS lights, speakers, and smartphone chargers, which have surpassed 200 million users; and self-driving technology Apollo, which has 50 municipal licenses in China to test autonomous vehicles on open roads.

"If anyone starts to be able to generate meaningful revenue (in AI), we will be the first to achieve that goal," Li told analysts on a recent earnings call, while acknowledging that Apollo is at a "very early stage."[8] Raymond Feng, an analyst at market research company Pacific Epoch in Shanghai, predicts that Baidu will start making money on driverless vehicle technology by 2020, providing improved AI services to vehicle manufacturers and drivers.

Baidu's core search business commands nearly three-quarters of China's search market and is reflected by the hulky block-like structure and hidden, inward office layout. Its logo is the paw print of a bear—a hunter's mark. But Baidu's lead is being chased by others in the BAT league. WeChat has added search features, while Alibaba is a backer of mobile search engine Shenma and Tencent-funded, NYSE-listed search company Sogou.

The most serious threat to Baidu's search kingdom is a potential reentry of Google to China. Baidu beat Google before, back in the late 1990s when Google China president Kai-Fu Lee was at the helm and making an all-out effort to win the market despite many challenges. When Google founders Larry Page and Sergey

Brin withdrew Google from China in 2010 in an uphill battle with Baidu and concerns about Chinese web censorship, Baidu promptly claimed two-thirds of the China search market. Baidu leader Li, a search expert who spent his early US career focused on perfecting web queries and links, told me then how Baidu won: a superior search engine in Mandarin and localized features such as community chat to handle queries. Baidu also made it in China by initially copying Google's paid search business model and, later, keyword-based marketing services and pay-for-performance online advertising.

As word was leaking out in midsummer 2018 about a possible reentry by Google, Li wrote a defensive message on his WeChat account, "If Google returns to China, we are very confident we can PK (or video game slang for player-kill) and win again."

But winning that battle is no sure bet. An internet poll on China's Twitter-like social media site Weibo shortly after the Google reentry news broke suggested that 86 percent of users would pick Google and its global reach over China-focused Baidu.

The A of the BAT

Alibaba's House of Cards

Alibaba's large shopping festival held annually on November 11 is like a gala celebration. Shoppers go mad purchasing online at discount prices, merchants unleash unbeatable promotions, and singers and dancers perform onstage at the huge Mercedes-Benz stadium along Shanghai's Huangpu River while a huge digital backdrop keeps score on incoming sales. This daylong online shopping fest, dubbed 11.11 or Singles Day, is a much larger version of the major US shopping holidays Black Friday and Cyber Monday. I was at the Singles Day festival in 2018, which Alibaba popularized as a marketing concept a decade ago, and got a behind-the-scenes look.

Alibaba's shopping promotion in 2018 hauled in a record $30.8 billion in merchandise sales, 27 percent more than the year before. That beats the $13.9 billion that Black Friday and Cyber Monday rang up in 2018 and the $3.5 billion in merchandise sales for Amazon's Prime Day shopping event for members.[9]

The e-commerce giant that Jack Ma and a team of 17 cofounders started in his Hangzhou apartment in 1999 has supersized, from a business-to-business online trading platform into a conglomerate of payments, logistics, and internet services that are disrupting banking, deliveries, and retailing. The headquarters today is a contemporary, sprawling campus of well-designed buildings and sculptures, in the city known for scenic West Lake—an upgrade from the drab office where I first interviewed Ma in 2006.

Then, Alibaba's original e-commerce site for small businesses in China and its online auction–style copy of eBay, Taobao, were just beginning to take off. But in 2010, when Alibaba beat eBay's then CEO Meg Whitman thanks to promotional stunts that got press, free customer listings, easy returns, and the dynamic leadership of Ma, suddenly the world spotlight began to shine on Alibaba. Taobao today counts around 700 million mobile users monthly. Its house of brands—Alipay payments, Alicloud cloud computing services, Alimama marketing platform, Aliwangwang instant messaging for customer negotiations, and its PR-ish news hub, Alizila—have taken it to the extreme in naming, like the Apple family of iPhone, iPad, and iTunes.

Ma broadened his view of the world as an English translator for tourists in Hangzhou. He loves to talk up Alibaba, its mission to help small businesses selling globally and ambition to last more than 100 years. A colorful character, he's dressed up in a Michael Jackson rock-star outfit and sung and danced to *The Lion King* theme song at past AliFest galas I've attended in Hangzhou. Certainly, Alibaba and its master know how to make a promotional splash.

Alibaba's high-profile Ma—never one to be shy—made headlines as one of the first global business executives to visit Donald Trump post-election at Trump Tower, promising to help create American jobs by helping small businesses sell to China on Alibaba e-commerce sites. I listened as he promoted Alibaba from the stage of a packed Waldorf Astoria ballroom before members of the elite Economic Club of New York. He's rubbed shoulders with who's who at the World Economic Forum at Davos—and hosted Bill Clinton, Arnold Schwarzenegger, and basketball star Kobe Bryant at previous AliFests where I've been one of thousands in the audience. With his gift for gab—in perfect English—Ma is now universally recognized by taxi drivers, teachers, brokers, and shop owners. This would have been hard to imagine when I first interviewed the stick-thin Ma in Hangzhou in 2006. Or when Alibaba venture investor and founding partner Tina Ju of Kleiner Perkins China told me how Alibaba almost went bankrupt after overexpanding in the dotcom boom and bust, only to be rescued by rigorously cutting the burn rate with layoffs and returning to China from overseas markets—plus the spiritual leadership of Ma and managerial knowhow of fixer and former General Electric executive Savio Kwan.

Alibaba Builds Anthills

A major component of the Alibaba vast empire is fintech giant Ant Financial, with its black, ant-shaped logo, a throwback to Ma's first Chinese startup, which he described as an ant up against an elephant. The elephant was Chinese government-owned China Telecom, which paid Ma $185,000 in a joint venture that ultimately killed his little startup. Ma once told me that you "use your brain when you're small," and no Chinese entrepreneur has managed to look bigger.

If taking over Yahoo! in China and beating eBay weren't big enough accomplishments, his move into fintech with Ant Financial is the one to watch. Ma spun Alipay out of Alibaba in a controversial move in 2011 that sparked a dispute with major shareholders Yahoo! and SoftBank over corporate governance standards and how compensation for the loss of Alipay would be made. Ma claimed the spin-off was necessary because of new Chinese government regulations that prohibit foreign ownership structures for payment services. An agreement was reached a year later with Yahoo! and SoftBank, guaranteeing they would get a share of the financial rewards in case the spun-off entity went public or got acquired. Shortly after Alibaba scored its mega IPO in New York in 2014, Alipay's financial services business was rebranded Ant Financial in a new push into financial services, and then in 2018, Alibaba crawled back in, buying a 33 percent stake in Ant Financial. Alibaba's fintech affiliate is shaking up the financial sector with internet technology and big data for wealth management, mobile payments, insurance, microloans, money market funds, and blockchain for cryptocurrencies. Its money market fund, Yu'e Bao, promising returns of more than 4 percent, became the world's largest fund in just four years after its 2013 launch, with $211 billion in assets and 370 million account holders, who needed only 15 cents to open an account. The fund's assets have since downsized to $168 billion following pressure by Chinese regulators and concerns of systemic liquidity risks to the entire banking market.[10]

Ant Financial made big news again when it hauled in the largest-ever single fund-raising by a private company: an eye-popping $14 billion investment in 2018 at a valuation of about $150 billion from US private equity firms Carlyle Group, Silver Lake Partners, Warburg Pincus, and General Atlantic, as well as Singaporean sovereign wealth fund GIC. In a sign of the intense competition in China and the power of Alibaba, investors in that round for its fintech affiliate had to commit to not making further investments in rival companies controlled by Chinese tech leaders Tencent, JD.com, and Meituan.[11]

Ant Financial could go public soon, and, if so, this ant will be riding on the coattails of Alibaba's record $230 billion valuation in its IPO on the NYSE.

The T of the BAT

Game on Tencent

At Tencent's vertical corporate campus in Shenzhen on a Friday late afternoon, a bunch of youngsters are hanging out on the steps and the entranceway. Banners are streaming, music is playing over blaring speakers, and loud laughs and chats are hard to miss. It looks like a giant party. Tencent's youth culture permeates its workplace and spurs new disruptive ideas. Several floors of its Shenzhen pair of towers are turned over to employees as a dining hall, a training center, and a fitness center equipped with running tracks and a basketball court.

Tencent derives its name from a fusion of the Chinese characters *Teng* and *Xun*, which mean "galloping fast information."[12] Tencent has lived up to that name though its brand mascot is a winking penguin wearing a red scarf. Best known for its Swiss Army–like WeChat messaging app in China, Tencent is also the world's largest video gaming company, bigger than Sony, Activision Blizzard, and Nintendo. Tencent's entertainment and communications colossus could be compared to Disney and Time Warner but more digital; more diversified in video, music, games, social networking, and content; and decidedly more Chinese. Tencent has built its tech base from southern China over the past two decades far from Los Angeles and New York.

Founded in 1988, Tencent has grown into a global technology leader by smart acquisitions, startup investments, and organic growth. Critics say Tencent relies too heavily on acquisitions and crushes startups. Keeping a creative spark in such a large company is a challenge—and it's trying. After TikTok beat the social conglomerate by several years with the fast uptake of its 15-second mini video clips, Tencent revived its own version, Weishi, in 2018.

Tencent has been on a highly profitable growth streak of nearly 10 years, reaping one-third of revenues from games and one-quarter from social networking and digital content. Tencent commands a 52 percent market share of China's $27.5 billion gaming market, the world's largest at 25 percent of the total. Tencent has a natural advantage, since China can sell games to the United States but American companies need to form joint ventures in China with the Chinese and share revenue with them.[13]

Tencent owns such blockbuster titles as *Honour of Kings,* a multiplayer role-playing game with 200 million players and nearly $2 billion in 2018 revenues. But Tencent's gaming business hit a major snag in mid-2018 when Chinese regulators clamped down on addictive and violent games and froze approval of new online games. When kids were banned from playing more than one hour per day or after 9:00 p.m. on video games, Tencent introduced a feature that could monitor playing time with facial recognition technology.

To keep the momentum going corporate-wide and help it stay competitive on many fronts, Tencent has stepped up investments in innovative technologies, zeroing in on its strongholds in mobile payments, social networks, and digital content and entertainment.

Across its online media platforms, Tencent is broadening a portfolio of digital music, video, and books as a content war heats up. Tencent saw digital content subscriptions grow by 50 percent to more than 100 million in 2018, a turning point and uptick for a relatively new source of revenue.

Tencent has been particularly busy shuffling a growing library of digital content. Tencent recently spun off Tencent Music Entertainment and China Literature, which each went public in Hong Kong and raised a total of $1.1 billion. They are profitable, rapidly growing businesses that expand Tencent's footprint and generate revenues from subscription fees, premium content, and online advertising.[14]

Tencent Music Entertainment, China's top music streaming service, offers up karaoke, live concerts, DJ mixes, sing-offs, and song recommendations. Though compared with Spotify, Tencent Music's 800 million monthly users outsize the Swedish company's listeners by more than four times. But for paid subscribers, Spotify's 83 million beat Tencent Music by three times. No matter—Tencent Music is profitable but Spotify isn't. Tencent relies more on pioneering virtual gifts and social entertainment services such as concert tickets for WeSing livestreaming of karaoke, rather than paid subscriptions or advertising to support its four music apps. In 2018, Tencent Music Entertainment revenues grew 73 percent to $2.76 billion and net profits reached $267 million.

Tencent's China Literature has evolved as the nation's largest e-book publisher, a source of content that can be adapted for movies and videos. China Literature is a budding force in film and TV through a recent $2.25 billion acquisition of powerhouse Chinese studio New Classics Media. China Literature innovatively makes personalized recommendations to online readers based on data analytics and advanced algorithms and leverages built-in social features to keep readers glued to the screen paying for premium online content distributed across Tencent's sprawling internet services, WeChat, QQ, and Tencent News. In 2018, China Literature chalked up a 23 percent revenue increase to $734 million and a 64 percent net profit gain to $133 million.

Another bright spot is Tencent's Netflix-styled and fully owned Tencent Video, China's largest streaming service, which has 89 million subscribers and is racing with Baidu's part-owned iQiyi at 87 million. Neither comes close to matching Netflix, which is closing in on 140 million subscribers globally.

Not much in the digital world can compare with the embrace of Tencent's flagship WeChat, or Weixin, social networking service, which counts more than 1 billion users who regularly spend more than an hour daily within the closed communication circles of the app. WeChat has shown its innovation stripes by evolving from messaging into an all-in-one superapp for e-commerce and payments, a trend in China tech with all titans racing to package in the most services.

"Chinese apps are more advanced in content, social networking, and commerce," observes Hans Tung, managing partner at GGV Capital. "The rise of 'super apps,' such as WeChat, Meituan, Ele.me, and Didi, have spawned a different model of app design versus Silicon Valley. China is showing the rest of the developing world that there is a different, and arguably better-suited, approach to scale locally: by becoming an ecosystem that can bundle different functions in one 'super app' for their users."

WeChat is getting good traction through its innovative new mini-programs, which bundle in shopping, gaming, and lifestyle services—and puts Tencent directly in Alibaba e-commerce turf. WeChat mini-programs reached the 1 million mark—half the size of the Apple App Store[15]—and 200 million daily users in less than two years of introduction in January 2017. The mini-shops across 200 service sectors blend content, ad units, and commerce seamlessly with links to sales products that can only be accessed within WeChat. It's a new concept from China, a monetization source that has originated in China's vast mobile universe.

American marketers seeking to reach Chinese customers see the appeal. Tesla has used mini-programs for users to schedule a test drive, find a charging station, and share their experience. Walmart launched its own scan-and-pay app within WeChat mini-programs.

"WeChat is becoming a default operating system and is fueling another wave of growth. Apps are being built on top of WeChat now," observes venture capitalist Tung.

On another front, WeChat is pushing into China's still-developing enterprise market with Qiye Weixin, an app launched in 2016 that offers office productivity software. Managers of US companies with operations in China often rely on WeChat for internal communications with their China peers using group chats and one-on-one texting and video calls. New York–based education startup Yoli has gone so far as to weave its English-learning apps connecting tutors and students into WeChat's platform rather than building its own app.

WeChat's brightest days could be fading since there is always something new in China's mobile universe. To wit, fired-up competitor ByteDance has launched a video chat app with built-in private messaging, Duoshan, to take on WeChat. The growth of WeChat is slowing, naturally, since it already reaches deep within China. Going global probably won't work. WhatsApp, owned by Facebook, is already entrenched in many markets with 1.5 billion users globally. In India's large market, WhatsApp is the mobile app to have. Persuading people outside China-centric circles to try WeChat is a stretch when it's difficult to initially grasp how it's better that texting. I'm okay with using WeChat personally, but my group, Silicon Dragon, has had mixed results with a WeChat official publishing account. The account was registered in the United States and couldn't be accessed in China, and when the account was shifted to China, instructions were in Chinese, all in all too challenging and troublesome for me to post content regularly.

The Up-and-Comers

Since the BAT got started in the first era of China's internet, a new group of net players has popped up in China that is focused on the mobile internet. This group is so potent that it has earned its own acronym: the TMD. The *T* is for AI-powered news aggregator Toutiao (pronounced *two ti ow*) and video app TikTok. The *M* is for food delivery and services app Meituan Dianping or the shortened Meituan (pronounced *may tia qan*), which was launched by the greatest cloner of China internet time, Wang Xing, who cloned Facebook, Twitter, Friendster, and Groupon. The *D* is for ride-sharing service

Didi Chuxing, or simply "dee dee," which absorbed Uber. I've added an *X* for Xiaomi (pronounced *shao me*), the world's fourth largest smartphone maker, from serial entrepreneur Lei Jun, who worships the legendary Steve Jobs.

Alibaba and Tencent at the head of the pack, plus Baidu not far behind, remain the epicenter of China tech—so far. That's pretty remarkable considering the sorry stories of Renren and Dangdang, two well-funded and -positioned Chinese peers that sprang up around the same time as the BAT. Both Renren and Dangdang were launched by US-educated and experienced returnees to China and backed by top-tier Silicon Valley venture shops. But both failed to live up to their potential in China's fiercely competitive digital commerce and communications market. They erred in imitating rather than innovating, and both lost focus by trying to do too many business lines at once.

China's once high-flying Facebook equivalent, Renren, fell after mainly copying Western models while Stanford-educated maverick founder Joe Chen put his attention on building an investment holding company of emerging tech startups—making good on his promise to "land grab everywhere," as he once explained his strategy to me. In 2018, Chen pulled off the masterful feat of selling Renren's failing social networking business and spinning off 44 of Renren's portfolio company investments, including well-known social finance business SoFi in a controversial deal to his privately held Oak Pacific Interactive, controlled by Chen and his Renren cofounders. Renren is now left with a used car sales platform in China, a trucking app in the United States, a (software-as-a-service) business for the US real estate market—and a sagging stock price on the NYSE.

Similarly, China's once promising Amazon-like book retailer, Dangdang, and its cofounding wife-and-husband team, Peggy YuYu and Li Guoqing, faltered. Looking to one-up Amazon CEO Jeff Bezos in China, YuYu used her Wall Street smarts to expand

Dangdang (sounds like a cash register) into selling apparel, toys, and linens and list the company on NYSE in 2010. But she ended up taking her e-commerce offspring private in 2016 following a wave of Chinese companies heading home for higher valuations. Back in China, Dangdang shed money-losing online merchandise categories to focus on books, and opened brick-and-mortar stores, returning to its roots after a planned $1.2 billion acquisition by Chinese conglomerate HNA Group fell through.

Global Reach?

For all their firepower, China's BAT companies are only starting to exert their power internationally, and it's slow going. International as a percentage of total revenues ranges on the low end for Baidu at 1 percent to Tencent at 5 percent and Alibaba at 11 percent. Going for a higher profile internationally, Alibaba made a splash with its first corporate ad push outside China as a major sponsor of the 2018 Winter Olympics in South Korea. Their American counterparts are vastly more international: Facebook and Google get about half their revenues from overseas while Amazon derives one-third from international markets.

Getting into US markets presents some unique challenges for China's BAT. Baidu's search engine works in the Chinese language. Its efforts to enter Japan in 2007 with a Japanese-language search engine, which has similar characters to Chinese, ultimately lost out to stronger international rivals Google and Yahoo!, leaving Baidu to exit the market in 2015.

Jack Ma's goal is to derive half of Alibaba sales from outside China, but it's been struggling to gain a foothold in the United States. Ma made several headline-making moves, like his promise to President Trump to create more American jobs and his Detroit fair to attract smaller businesses to sign up with Alibaba, but the US-China trade war is a negative.

Tencent's WeChat, for all its popularity in China, just doesn't move the needle much in the United States. Only about 2 percent of US internet users access WeChat once every few days—although the majority of my Chinese American business contacts in the United States and in China vouch for its usefulness. WeChat takes some getting used to: no stored history of texts that are easily accessible is one issue. And while US users of WeChat can chat and text, its popular payment features don't work well in the United States. WeChat Pay and its distant cousin Alipay are only now getting around restrictions on their use outside China by aligning with multinational payment services PayPal and Stripe. These Chinese payment apps have been used mainly by Chinese executives, tourists, or students holding a Chinese bank account and a national ID card.

US Markets a Tough Nut

Don't expect China's leading tech innovators to go mainstream in the United States anytime soon. Standing in the way are cultural differences, lack of brand recognition, and government regulations. Fears that Chinese products are not safe to use is another deterrent. "Going global is going to be a long, tough haul for these Chinese brands," says China expert Ann Lee, who is CEO of new technology investment consortium Coterie in New York. "There is a lack of trust of Chinese brands in the US. It's a psychological, emotional factor."[16]

Chinese Government Crackdown

Perhaps the biggest risk to the staying power of Baidu, Alibaba, and Tencent is a crackdown by the Chinese government on their considerable social impact and monopolistic power. "The biggest thing to watch is how the Chinese government will impact these companies if the government identifies them as a risk to their power," says Lee, an authority of China's economy and author of *Will China's Economy*

Collapse?[17] "If the government hurts them in their home market, they could lose credibility. The government could shut them down."

Jack Ma has recently said he is a member of the Communist Party. Making good with government in China is not a bad idea.

The Chinese government has put pressure on China's internet giants to clean up questionable content. Baidu, Alibaba, and Tencent have all pledged to step up scrutiny of videos uploaded by users and are spending more time sanitizing politically sensitive content, pornography, off-color humor, and excessive celebrity gossip. Tencent has limited video game time for underage players (one hour daily for children under 12 years old and two hours for teens) of its megahit smartphone game *Honour of Kings*—and uses facial recognition technology to detect minors who are playing. Tencent also recently started enforcing real-name verification for the game—and planned to expand this practice to its entire game lineup. In a patriotic move, Tencent rolled out a gaming app for players to compete in applause of remarks made by Chinese president Xi Jinping in his speech at the Communist Party's congress in November 2017.

Keeping Data Private

Similar to the US tech leaders, China's technology giants are also struggling with data privacy issues and monitoring of personal information. Following public outrage over a social credit service that automatically enrolled users and gave access to other companies and third parties about their personal income, savings, and shopping expenses, Alipay management publicly apologized and took down the feature early in 2018. In an ominous sign of what the future could hold for China's tech superpowers, media reports have sometimes surfaced that Beijing has discussed taking small equity stakes to have a management hand in China's social media giants, including the Twitter-like Weibo and Alibaba-owned video site Youku Tudou.

The Big Three's Future

What's next for Baidu, Alibaba, and Tencent? China's BAT is at the forefront of the country's tech revolution. They own large chunks of China's tech economy from maturing businesses in search, commerce, and communications, and they're moving fast into cutting-edge technologies in AI, robotics, and fintech. Their acquisitive push outside China into the United States and Southeast Asia is giving China's three tech titans power and influence around the world. Someday soon, Baidu, Alibaba, and Tencent may be as well known on Main Streets and Wall Streets around the world as they are in Beijing, Shanghai, and Shenzhen. Whether they will ever have the international recognition of Facebook, Amazon, Netflix, and Google globally is questionable, but it's a topic on the radar that wouldn't have been considered 10 years ago. The dragon has awakened.

CHAPTER 2

TECH GIANTS TAKE A BITE AND BULK UP

China's technology giants are powering up in a world race to buy into and acquire cutting-edge digital startups and own it all, despite growing tensions over a US-China trade war and tech leadership of tomorrow.

Jack Ma, now in his mid-fifties, famously said a few years ago that he is too old to run an internet company. He's stepping down as chairman and turning over the reins to CEO Daniel Zhang, a 12-year veteran of the company who originated Alibaba's highly successful 11.11 shopping festival. Ma has been acting more and more as a figurehead, ceding power to the next leaders to ensure a smooth transition. It's a tricky time, with a US-China trade and tech war impacting American merchants selling to China and Chinese investment in the United States, a new digitalized retailing environment requiring constant investment, and intense competition with domestic rival Tencent. For Chinese tech companies, the immediate impulse is to power up as a shield to protect their business and knock out interlopers.

I met two of Alibaba's top execs at a dinner in Shanghai hosted for key opinion leaders during the 11.11 shopping festivities in

November 2018. Michael Evans, who previously ran Goldman Sachs' Asia operations advising Ma as a banker on power-building deals, is president, in charge of spearheading growth in international markets. The polished Princeton University graduate and experienced Wall Street executive is equally at home on Fifth Avenue or Nanjing Road, and is pumped up about the prospect of making Alibaba more global. Alibaba cofounder Joe Tsai, a former private equity investment manager with Sweden's Wallenberg family–controlled Investor AB, has been elevated from chief financial officer to executive vice chairman, leading strategic acquisitions and investments. The Yale Law School graduate and sports fanatic (he owns the San Diego lacrosse franchise and part-owns the Brooklyn Nets basketball team) is the only one of Alibaba's founding team with a Western education, and he set up the Chinese company's financial and legal structure. From our conversations, it's clear that conquering the e-commerce world means that mergers and acquisitions and venture investments in China and globally will be a more important playing card for Alibaba's future. Ma was not present for dinner; he was dining with Chinese government officials on a Saturday night.

Bigger Is Better

China's gigantic tech companies have followed the conventional business wisdom of bigger is better. They have gotten big fast in what matters, investing strategically as a base to build and maintain power. Owning more and more is an overriding strategy of the high-tech titans. They have reached into large tech-centric economic sectors in health care, education, finance, and biotech. "They have started investing broadly in several sectors both domestically and abroad," observes Chris Evdemon, a venture partner at Sinovation Ventures who has worked in both Beijing and Silicon Valley. "There is no aspect of technology that is not of interest to the BAT."

Their "supersize it" strategies outdo the American FANGs, which have acquired businesses closer to their home base and core sectors, such as Instagram, YouTube, WhatsApp, Messenger, Waze, Alexa, Zappos, and Twitch.

For China's top tech companies, it's about "broadening their footprint and influence" by ploughing excess cash back into investments and building "vast constellations of satellites," notes Sequoia Capital partner Mike Moritz. This acquisition-heavy approach differs from the largest US tech companies, which spend far more on stock buybacks and dividends, he points out. "Uber, Airbnb, and SpaceX may be hogging the limelight, but the undisputed gold medal leaders are the Chinese," opines Moritz, noting the scale and acquisitiveness of China's tech titans.[1]

The BAT's Buying Binges

For several years, China's big three tech giants have been on a US buying binge, going after the gems. From 2010 to 2018, Baidu, Alibaba, and Tencent inked 227 tech deals worth $33.5 billion, or two-thirds of the overall $51.4 billion Chinese investment in US tech. The most acquisitive by far is Tencent with 146 deals and $25.7 billion of investment, followed by Alibaba with 51 deals and its part-owned Alipay with 2 deals and $3.7 billion in volume, and Baidu with 28 tech investments at $4.1 billion.[2]

China's dragons have teamed up with top-tier US-based venture firms Mayfield and New Enterprise Associates, private equity firms General Atlantic and Carlyle Group, corporate strategic investors General Motors and Warner Brothers, and Japan's acquisitive SoftBank. They've invested in US ride-hailing leaders Uber and Lyft, electric-carmaker Tesla, and augmented reality innovator Magic Leap.

These Chinese tech titans have taken their cues directly from Silicon Valley venture capitalists. They've scoured the Valley for promising startups and based their operations not far from Menlo

Park's storied Sand Hill Road firms that backed winners Google, Facebook, and eBay.

Tencent opened an office in a converted church in tech-wealthy Palo Alto, home to Stanford University, and has expanded nearby to a much larger California base. Alibaba keeps an office in San Mateo on California byway El Camino Real, in sight of venture capitalist Tim Draper's entrepreneurial school Draper University. Baidu has established two research and AI labs in high-tech Sunnyvale, Silicon Valley central.

China's tech titans have co-invested in the United States with many hot-shot Silicon Valley venture firms, including influential Andreessen Horowitz, whose lead partners founded once-dominant web browser Netscape, which was acquired by AOL for $4.2 billion. In their quest for California's deep tech riches, China's big three investors have often been regarded in Silicon Valley as premium buyers who pay more and help to provide access to huge China markets. "They can have their pick of the litter," said David Williams, founder and CEO of Palo Alto–based investment banking firm Williams Capital Advisors. "They have a VC-like perspective and are fast-moving."

Soft Power Grab: Hollywood

China's tech giants haven't skipped Hollywood in their love story with California. Alibaba's Jack Ma, who once starred in a kung fu movie alongside Jet Li, teamed up with Steven Spielberg's film group, Amblin Entertainment, to produce and bring films to China, and he formed Alibaba Pictures as a Hollywood-to-China entertainment hub. Tencent backed Burbank studio STX Entertainment, partnered with Dick Clark Productions to bring the Golden Globes and Billboard Music Awards to China, and invested in Hollywood blockbuster *Wonder Woman*. And Tencent cofinanced the production of

2019's *Terminator: Dark Fate* and handled its distribution into Asia. China's Dalian Wanda Group acquired movie theater operator AMC and purchased film production company Legendary Pictures. These moves into Tinseltown culture represent China's bid for soft power in America's most iconic field and an eagerness to create its own Hollywood-like studios and productions—*Chollywood?*—in the country's huge and dynamic movie market. Dalian Wanda built the largest movie studio in the world in the northern port city of Qingdao and has been luring foreign producers to shoot films in the sprawling complex. But a US-China studio coproduction of big-budget action film *The Great Wall,* starring Matt Damon, was a box-office flop.

Currents have been flowing for cross-border investments between the United States and China for several years. But the waves are subsiding as recent regulatory pressures and financial crunches cramp deal makers on both sides.

US regulatory controls have recently tightened to protect America's competitive edge by restricting foreign investment in strategic US technologies that may pose an economic and security risk. China has eased up on making more deals in the United States. In 2018, Chinese companies made 80 M&A and private placement transactions with US tech companies—slightly less than 89 the year before—but deal volume plunged to $2.2 billion from $10.5 billion in 2017. In the peak year of 2016, there were 107 Chinese deals in the United States, totaling $18.7 billion.[3]

The increased scrutiny over China acquisitions and investments in US tech companies also prompted the three BAT companies to hit the pause button. In 2018, the threesome made 31 US tech transactions, slightly more deals than a year earlier but at smaller volumes and in highly strategic businesses in comparison to trophy deals in the past.[4]

Meanwhile, China direct investment in the United States

dropped to $4.8 billion in 2018 from $29 billion in 2017—the lowest level in seven years after a 2016 peak of $46 billion invested in 162 deals.[5,6] Few new deals are pending, a five-year low.

From Beijing, a crackdown over highly leveraged deals in the United States—largely by Chinese conglomerates in US tech, real estate, wineries, and Hollywood—has put the skids on cross-border deal making. Chinese companies shed $13 billion of US assets in 2018 and another $20 billion is pending, as Beijing pressures companies to prioritize cutting debt over global expansion. Dalian Wanda sold its Beverly Hills condo-hotel project, while HNA let go of its stake in Hilton Worldwide and ownership of a Manhattan office building near Trump Tower. The Chinese government may offload the iconic Waldorf Astoria Hotel it took over in February 2018 from China's debt-ridden Anbang Insurance Group. Anbang had purchased the Waldorf for $1.9 billion from the Blackstone Group in 2014 and was in the midst of a three-year, $2 billion gut renovation of the luxury hotel.

These tremors carried over to Hollywood. A $1 billion deal by Chinese real estate and entertainment conglomerate Dalian Wanda to acquire Dick Clark Productions, producer of the Golden Globes and American Music Awards, collapsed. Regulatory pressures, as well as payment issues from Wanda's side, were to blame. Paramount Pictures' $1 billion film financing deal with China's Huahua Media also fell apart. In a related entertainment and distribution deal meant to capitalize on the trend toward livestreaming, a $2 billion agreement by Chinese tech and entertainment conglomerate LeEco to acquire Los Angeles–based TV maker Vizio was called off over a cash crunch and regulatory issues. Deals worth an estimated $8 billion were abandoned in 2017 due to unresolvable concerns over foreign investment in the United States.[7]

China deals in the United States have faced tougher approvals in the Trump administration, but many China-to-US technology buys are still getting the go-ahead if they are structured as investments rather than acquisitions, outside of critical technology sectors.[8,9] Under Trump, the recently enacted Foreign Investment Risk Review Modernization Act expanded reviews of the Committee on Foreign Investment (known as CFIUS) and targeted acquisitions, minority deals, and venture investments into critical technologies such as semiconductors, autonomous driving, or potential military applications. A takeover of San Diego–based chip maker Qualcomm by Singapore rival Broadcom was shot down over potential security risks in 2018, as an example of the tightening over foreign ownership.

Possible additional restrictions may be coming as part of an update of the US Trade Representative's "Section 301" investigations into China's intellectual property and technology transfer practices.

The heightened regulations and uncertainty over approvals are causing China tech titans that previously singled out the United States to turn to other strong technology centers, such as Israel and to the booming Southeast Asian region. Alibaba, which was caught in the headwinds, is pivoting after a block on its affiliate Ant Financial from acquiring Dallas-based money transfer service MoneyGram for $1.2 billion in 2017. US regulators had raised issues about security and privacy risks for stateside users. To push the deal past national security concerns, Ant Financial had promised to keep the MoneyGram personal financial information secure by storing the data on servers in the United States. But the deal wasn't approved and Alibaba paid a $30 million termination fee to MoneyGram. Following that rejection, Alibaba has made only a few tech deals in America, and those were highly strategic, smaller ones, such as an acquisition of New York–based social shopping marketplace OpenSky.

"Given the current environment, it remains to be seen what investments Alibaba will or can make in the US," says Hans Tung at GGV Capital, a venture firm that was an early backer of Alibaba. Going "where they're more welcomed, such as Southeast Asia or India seems to make more sense."

Pivot to Israel

Alibaba's Ma has turned his kung fu–like skills to seeking and funding startups in the "Startup Nation" of Israel. On his first trip to Israel in May 2018, he led a delegation of 35 Alibaba executives to visit investors and check out startups in Israel's stronghold of cybersecurity as well as augmented reality, online gaming, QR codes, and AI. Alibaba promptly invested $26 million in big data company SQream Technologies, co-invested $40 million in mass transit software startup Optibus, and added to its $30 million co-investment in safe driving technology startup Nexar. These deals were on top of its first Israeli deal, an acquisition of personalized QR code designer Visualead in 2017 to establish a Tel Aviv research and development center. China to Israel deals are a growing trend, marrying capital and market potential. In a further move in Israel, Alibaba expanded its high-tech research lab DAMO (discovery, adventure, momentum, and outlook) Academy to Israel.

Southeast Asia: The Next China

China's forward investment march is leading to Southeast Asia and the region's high-potential, populous, and digitally savvy markets. This is a trend that has earned the label "Chuhai," which is actually a canned alcoholic drink from Japan but is also a Chinese term being used to describe the phenomenon of Chinese entrepreneurs targeting emerging markets outside of China as the mobile internet market at home reaches saturation.[10] China has plowed more than two-thirds of its overseas tech investment in recent years to Asia.[11] Led by China's three leading tech titans, big sums of money are going into in e-commerce, search, and ride-hailing startups in Singapore, Vietnam, Indonesia, Malaysia, and India. These regional deals closely parallel their power base in China and investments in Chinese startups. The potential is huge: Asian startups typically lag China in development by at least five years. The gap creates a good opportunity for Chinese investors to profit from investing in next-generation tech stars in Asia, a topic I wrote about in my forward-looking book *Startup Asia*.[12]

China is reaping the bonus of an early start in Southeast Asia and is making American companies seem slow and clumsy. Take what's happened with Uber. After Uber was overtaken by local Chinese rival Didi in 2016, the American ride-hailing leader then sold out to its chief Asian competitor, Grab, Southeast Asia's dominant ride-share company. At Amazon, founder and CEO Jeff Bezos made a big fuss about committing $5 billion to pursue India's tremendous potential. Indian newspapers ran front-page headlines of Bezos when he kick-started Amazon in India by parading into Mumbai on the back of a colorful truck, with a $2 billion investment check in tow. But Walmart beat him to a big prize in India by buying Indian online retailer leader Flipkart in a $16 billion deal in 2018. See table 2-1.

Table 2-1

The BAT and FANGs Target Southeast Asia

RIDE-HAILING

Tencent
Led $1.1 billion co-investment in Ola in India, 2017
Led $1.2 billion co-investment in Go-Jek in Indonesia in 2017

US Brands
Uber was acquired by Grab Singapore, and Uber got a 27.5% stake in Grab in 2018
Google co-invested $1.2 billion in Go-Jek in Indonesia in 2017

E-COMMERCE

Alibaba, Ant Financial
Invested $4 billion in Lazada Group, Singapore, 2016–2018
Led two $1.1 billion co-investments in Tokopedia in Indonesia, 2017 and 2018
Co-Invested $1.3 billion in Paytm in India, 2015–2018

US Brands
Amazon invested $5 billion in India since 2014
Walmart spent $16 billion to acquire a 77% stake in Indian e-commerce leader
    Flipkart in 2018

Sources: Silicon Dragon research, S&P Global Intelligence, annual reports, news releases

To fortify its stronghold, Alibaba has paid big sums for chunks of Southeast Asian regional tech leaders, notably spending $4 billion for a controlling stake in Singapore-based e-commerce leader Lazada and co-investing a total of $2.2 billion in Indonesian mobile payment service Tokopedia. See table 2-2.

Tencent is also hunting in the region and has invested in fast-growth ride-hailing and e-commerce leaders in India and Indonesia plus startups in Vietnam and Thailand.

China's BAT Push Outward

Despite the growing frictions and challenges on the US-China trade and tech fronts, China tech companies are ambitiously pushing to go global in a winner-takes-all economy. The three Chinese high-tech titans are each pursuing investments beyond national borders and original business sectors. Let's look at the strategies of each of these titans, in order of their place in the BAT league.

Buffing Up Baidu

Baidu is betting its future squarely on diversifying beyond search and into artificial intelligence technologies for self-driving, smart

Table 2-2

Sampling of Alibaba and Tencent Investments in Southeast Asia

Company	Inv. Type	Inv. Amt	Market	Country	Year
Alibaba					
Tokopedia	Co-inv.	$1.1 billion	e-commerce	Indonesia	2018
Lazada	Inv.	$4 billion	e-commerce	Singapore	2016–2018
Daraz	Acq.	$200 million	online shopping	Pakistan	2018
Paytm	Inv.	$222 million	online payment	India	2017–2018
Tokopedia	Inv.	$1.1 billion	e-commerce	Indonesia	2017
Tencent					
Gaana	Lead Inv.	$115 million	music streaming	India	2018
Tiki	Stake	Und.	e-commerce	Vietnam	2018
Ola	Lead Co-inv.	$1.1 billion	ride hailing app	India	2017
Flipkart	Co-inv.	$1.4 billion	e-commerce	India	2017
Go-Jek	Co-inv.	$1.2 billion	scooter hailing	Indonesia	2017
Ookbee	Inv.	$19 million	digital content	Thailand	2017
Pomelo	Lead Co-inv.	$19 million	online fashion	Thailand	2017
Sanook	Acq.	Und.	web portal	Thailand	2016

* Note–Inv. is investment; Co-inv. is co-investment; Acq. is acquisition; Lead Inv. is lead investment; Lead Co-inv. is lead co-investment; Und. is undisclosed

Sources: Silicon Dragon research, S&P Global Intelligence, annual reports, news releases

transport and voice-assisted smart home devices. In a major restructuring of its China business a few years ago, Baidu ditched several costly peripheral, cash-burning online businesses in food delivery services, mobile games, online travel, web shopping, and health care that were losing a battle with hard-charging Tencent and Alibaba. Baidu's takeout delivery service, Waimai, was sold to rival startup Ele. me, now owned by Alibaba. Its smartphone app store 91 Wireless, which Baidu had acquired a few years earlier for $1.9 billion, was offloaded. Baidu also let go of its interest in money-losing online travel business Qunar to rival Ctrip for a stake in the consolidated business. The shuttered online shopping store Youa and two forays into medicine, including a Baidu Doctor mobile app, were also sloughed off. Still left was a small stake in Uber, which Baidu had gained through a $1.2 billion co-investment in Uber's China business intended to integrate the ride-hailing app into its popular mapping service. When Uber was overtaken by Chinese rival Didi, Baidu ended up with a tiny stake in Didi, which now faces its own challenges.

Not finished with rejiggering its business lines, Baidu spun off its video streaming business iQiyi (pronounced *eye-chee-yee*) in 2018, which then scored a New York IPO that pulled in $2.3 billion at a market valuation of $12.7 billion, one of the year's highest on US exchanges. Baidu's Movie-star-handsome founder, who is an internet celebrity in China, traveled to New York for the bell-ringing ceremony. Seeing him there brought back memories of Li proudly posing at Nasdaq's Times Square headquarters for Baidu's IPO in 2005, when Baidu raised $109 million, then seemingly a large sum for a Chinese tech company.

Determined to keep a lead in cutting-edge AI technology, Baidu budgeted $300 million for a second Silicon Valley research lab in 2017, supplementing its first in 2014, and the Beijing-based titan has set up an engineering office in Seattle to focus on autonomous driving and internet security. Baidu has pumped loads of capital into AI startups in the United States with technologies for deep learning, data

Table 2-3

Sampling of BAT Investments in US Technologies—2018

Company	Inv. Type	Inv. Amt	Market
Lunewave	Co-inv.	$5 million	self-driving sensors
Vesper Tech.	Co-inv.	$25 million	acoustic sensors
SalesHero	Co-inv.	$4.5 million	AI sales assistant
Sensoro	Co-inv.	Und.	IoT sensors

* Note–Inv. is investment; Co-inv. is co-investment; Und. is undisclosed

Sources: Silicon Dragon, S&P Global Market Intelligence

analytics, and computer vision. See table 2-3. "Having missed out on the social mobile and e-commerce waves of the past few years, Baidu is trying not to repeat the same mistake by going all in on AI, on all fronts," observes Evdemon of Sinovation Ventures, the Beijing-based venture capital firm headed by AI expert and investor Kai-Fu Lee.

Alibaba's New Investor-Hungry Team

From my discussions at the recent 11.11 festival with Alibaba leaders Tsai and Evans, it's clear that Alibaba considers investments to be equally important to organic growth. See table 2-4. In one recent power-packed move, Alibaba bought control of food delivery service Ele.me in a 2018 deal that valued the service at $9.5 billion, then merged it with its local commerce services entity Koubei and raised $3 billion for the combined business. Alibaba's muscular move counters Tencent-backed delivery and services app Meituan. Not that all transactions have worked out for Alibaba. One Chinese deal that looks like a loser is Alibaba's $1.5 billion investment in cash-burning, bike-sharing startup Ofo, once a star at the height of China's shared-bicycling craze. Ofo's key rival Mobike, backed by Tencent, was absorbed into Meituan as Meituan Bike in an acquisition.

Table 2-4

Alibaba Investments in US Tech Startups

Company	Inv. Type	Inv. Amt	Market	Year
Smartrac	Inv.	Und.	RFID, IoT	2018
OpenSky	Acq.	Und.	B2B e-commerce	2018
NVXL Technology	Inv.	$20 million	machine learning	2017
EyeVerify	Acq.	$100 million	security	2016
Snap	Inv.	$200 million	photo app	2015
Lyft	Co-inv.	$250 million	ride-sharing	2014
Quixey	Co-inv.	$110 million	mobile search	2013–15
Tango.me	Co-inv.	$280 million	messaging app	2014
Kabam	Inv.	$120 million	gaming	2014

Alibaba Investments in China Tech Startups

Company	Inv. Type	Inv. Amt	Market	Year
Cainiao	Lead Co-inv.	$1.4 billion	smart logistics	2018
Ele.me	Acq.	$9.5 billion	food delivery	2018
Ele.me / Koubei	Merger			2018
Koubei	Acq.	$1 billion	local commerce	2017
Xiaohongshu	Lead Co-inv.	$300 million	social e-commerce	2018
Ofo	Inv.	$866 million	bike sharing	2018
SenseTime	Inv.	$600 million	facial recognition	2018
Ofo	Inv.	$700 million	bike sharing	2017
Youku Tudou	Acq.	$4 billion	video sharing	2016
Weibo	Inv.	$720 million	micro-blogging	2016
AutoNavi	Acq.	$1.5 billion	digital mapping	2014

* Note—Inv. is investment; Co-inv. is co-investment; Acq. is acquisition; Lead Inv. is lead investment; Lead Co-inv. is lead co-investment; Und. is undisclosed

Sources: Silicon Dragon research, S&P Global Intelligence, annual reports, news releases

In the United States, Alibaba has had a mixed record of M&A deals. A $100 million acquisition of eye scan security startup EyeVerify in Kansas City was well planned and executed. EyeVerify became the global center of Alipay for mobile biometrics or eye scans to verify IDs for banking, mobile payments, and security. But a series of Silicon Valley–style startup deals that Alibaba made in the run-up to its IPO in 2014 ultimately failed either because of product fit mismatches with the China market or missed milestones: TangoME in mobile messaging, Kabam in gaming, and Quixey in mobile search. The big blow to Alibaba's US ambitions was the block on its Ant Financial deal to buy money transfer firm MoneyGram.

Tencent's Surround-Sound Strategy

At Tencent, its laser focus on strategic investments in diverse companies is seen as a tool to get ahead in frontier technologies such as connected cars and internet-facilitated health care. The investment outreach is also a shield against downturns from any further regulatory turmoil in the gaming sector. Tencent has made over 700 such investments and has a good track record. More than 100 of its investee companies have reached valuations exceeding $1 billion and 60 have gone public, one dozen since 2017. One recent win was Tencent's pre-IPO investment in China's next-generation titan Meituan, which yielded a gain of about $1.3 billion.

A core of Tencent's corporate culture is rapid-fire acquisitions and investments, a quicker way to realize results than internal innovation, which can take years to develop with an uncertain payoff—with some exceptions, like WeChat.

Tencent's strategy of doing everything loops in financial services: wealth management, insurance services, consumer loans, and WeChat Pay. Tencent hasn't skipped the artificial intelligence revolution either and has made as many as 25 investments in AI startups in China.

In the United States, many of Tencent's early investment deals were for high-profile American technology trophies Uber, Tesla, and Snap, but the action has settled to smaller strategic buys in startups across biotech, games, and robotics. See table 2-5. In early 2019, Tencent went off script and made the bold move of co-investing $300 million in US social news aggregator Reddit—a deal that sparked some user protests over expected censorship issues from having a Chinese company as an investor.

It's not easy to retain leadership across such diversified sectors. Could Tencent be dislodged as the Chinese king of social networking and video games? Not likely anytime soon.

Tencent CEO Ma and his handpicked president, Marvin Lau, a former Goldman Sachs banker with dual master's degrees from Stanford and Northwestern, wants to make sure that doesn't happen. Tencent recently launched a program to nurture young talent by committing to promote younger employees to one in five open positions. Now in its twentieth year, Tencent was restructured for the first time in six years to focus on business services such as cloud computing and payments. Tellingly, a technology committee was formed to strengthen research and development.

Games and More Game Buys

Acquiring and investing in games and more games have kept Tencent and its investment bankers and legal team very busy over the years. In 2018 alone, Tencent invested or became a majority owner in four of the five biggest gaming deals of 2018, among them $2 billion for Vivendi's stake in French video game developer Ubisoft as well as several smaller investments in Chinese gaming players.[13] In the United States, Tencent spent $400 million in 2015 to acquire Los Angeles–based Riot Games, operator of the highly popular PC game *League of Legends*. Tencent's biggest gaming deal was snapping up Supercell in Finland in 2016 for a whopping $8.6 billion. Tencent

Table 2-5

A Sampling of Tencent's Investments in US Tech Companies

Company	Inv. Type	Inv. Amt	Market	Year
Activision Blizzard	5% stake	$2.3 billion	interactive entertainment	2013
Epic Games	48% stake	$330 million	video game and software	2013
Fab.com	Co-inv.	$150 million	online home décor	2013
Riot Games	Acq.	$400 million	game developer	2015
Glu Mobile	Inv. 15%	$126 million	game developer	2015
Pocket Gems	Inv. (+2017)	$150 million	mobile video game	2015
Smule	Lead Inv.	$54 million	karaoke app	2017
Snap	12% stake	$2 billion	video-messaging app	2017
Uber	Co-inv.	$1.25 billion	ride hailing	2017
Tesla	5% stake	Und.	electric vehicle maker	2017
Grail	Joint Inv.	$900 million	cancer detection	2017
Essential Products	Inv.	$300 million	consumer electronics	2017
VoxelCloud	Lead Inv.	$15 million	medical AI	2017
Locus Bioscience	Co-inv.	$5 million	biotech	2017
Hammer & Chisel	Co-inv.	$150 million	game developer	2018
Capture Technologies	Co-inv.	$1 million	event data analytics	2018
Marble	Co-inv.	$10 million	robotic delivery	2018
Skydance Media	Inv.	Und.	film/VR	2018
Reddit	Co-inv.	$300 million	social news aggregator	2019

* Note—Inv. is investment; Co-inv. is co-investment, Acq. is acquisition; Lead Inv. is lead investment; Lead Co-inv. is lead co-investment; Und. is undisclosed

Sources: Silicon Dragon, S&P Global Market Intelligence, annual reports, news releases

also became a minority shareholder in 2012 of American video game company Epic Games, the studio behind gaming sensation *Fortnite*.

All was going along fine when Tencent's gaming business hit a speed bump in mid-2018: Chinese regulators froze approvals of new online gaming titles in a crackdown on addictive and violent content. Tencent profits dipped, sales growth of online games slowed, and share prices plunged. Longtime stockholder South African media and internet group Naspers reduced its holdings by 2 percent to 31 percent, earning a cool $10 billion from funding Tencent starting back in 2001. Things were looking up for Tencent in December 2018, when China's watchdog gave Tencent approval for a batch of new smartphone games. Tencent pulled through the year with a 24 percent revenue increase for mobile games but an 8 percent decrease for PC games.

Playing War with Alibaba

In its home base of China, Tencent's acquisitions style is like a warrior, fighting its key rivals on multiple fronts. Tencent has targeted Alibaba with killer buys in e-commerce, such as taking an 18.5 percent investment stake in social commerce disrupter and Nasdaq-listed Pinduoduo. It's also invested in video app Kuaishou, a competitor to TikTok's quick-start video app.

Earlier on, in 2013 and 2014, Tencent mainly took small stakes in then-hot Chinese web and search startups: $736 million for 20 percent of Craigslist-like 58.com; $448 million for 36.5 percent of search engine Sogou; $180 million for 15 percent of online real estate services platform Leju; and a double-digit stake in app maker Cheetah Mobile.

Is Tencent Failing to Realize Its Dream?

This long string of acquisitions and deals that Tencent has made has led some observers to conclude that this tech giant has lost its dream and passion for innovation.[14,15] "Tencent is ignoring the core competitiveness of a technology company that should come from product innovation," wrote tech blogger Pan Luan in a widely read essay that faulted the company for becoming an investor instead of an innovator in core areas.

But Tencent defends its diversification strategy, which has been likened to "sprinkling pepper" [on food]. Tencent Investment Partnership Manager Li Zhaohui in an interview explained the logic of its deal making: Tencent only invests in areas related to its core business of consumer internet, but Tencent is constantly entering new areas because of how the internet is expanding and converging across sectors.[16]

Can Tencent pull off another win as big as its internally developed WeChat and rev up its gaming businesses? It's really up to Tencent to lose. With its reach into communications and entertainment, Tencent's pinnacle position will be extremely hard to overtake—and certainly no American competitor can really try in China.

CHAPTER 3

GAINING FAST: CHINA'S NEXT TECH TITANS

The next group of up-and-comers is right behind China's
BAT and leading the future for smartphones that rival Apple,
internet-connected smart homes, superapps for speedy on-demand
takeout lunches, plus 15-second video thrills and AI-fed news.

XIAOMI: The Apple of the East

Chinese tech entrepreneur Lei Jun is sometimes called the Steve Jobs of Apple. An entrepreneur celebrity in China much as Jobs was in Silicon Valley, he launched China's smartphone maker Xiaomi in the spirit of Jobs and copied his products and style down to blue jeans and black T-shirt attire and stage presentations for new iPhones and iPads, even once teasing an introduction with the adopted line "Just one more thing." Xiaomi mobile phones that carry the brand name Mi (like the i in iPhone) have been called a less expensive imitation of the original, groundbreaking iPhone. Its retail stores resemble the minimalist design of Apple stores. A multibillionaire angel investor and serial entrepreneur, Lei readily acknowledges that he wanted to follow in Jobs' footsteps and achieve a certain cool factor after reading a book *Fire in the Valley*, about the early days of the personal computer industry. His admiration of Jobs shows how China has always looked up to Silicon Valley as the pinnacle. Of the Chinese tech entrepreneurs I've met, smartphone maker Lei is the closest to the legendary Jobs. And now, Apple is copying Xiaomi by integrating more revenue-producing, subscription-based entertainment and news content into iPhones—something that China's Xiaomi phones have had since day one.

The Chinese market demands hyperspeed and precise execution—and an eye on superfast growth before profitability. It can be a winner-takes-all market. A few other Chinese companies have emerged in this league besides Xiaomi with its cool smartphones. They include AI-powered news and video apps Toutiao and TikTok, superapp Meituan, and ride-hailing service Didi. This group, the TMD or Toutiao, Meituan, and Didi, echoes China's BAT companies. TMD is also slang in Chinese for a rude profanity, so I don't like to use the acronym much. Instead, let's call this group of next-generation BAT companies XTMD, adding the X for Xiaomi. These companies are mobile-centric for today's always-on young generation, leverage the latest technologies in AI and data analytics, and set the pace in innovation and scale. Their business models and features are often ahead of the West and sometimes are copied.

Toutiao uses machine learning to distribute personalized streams of aggregated content to online readers, a BuzzFeed with brains. TikTok spins out 15-second music video clips, akin to Snapchat's Lens Challenges launched two years after the Chinese app. Meituan is a one-stop superapp that rolls in Uber Eats, Kayak, Yelp, and Groupon in a range of services including food delivery, travel bookings, and movie tickets. The Uber of China, ride-hailing startup Didi Chuxing, actually beat Uber in China. Xiaomi is primarily known as a lower-cost, high-quality smartphone maker but makes most of its money from internet services and internet-connected gadgets.

These next-tier titans are at the forefront of breakthrough homegrown Chinese tech companies and have raised venture capital at lofty valuations in the billions of dollars stratosphere. Two, Xiaomi and Meituan Dianping, have already gone public, in 2018. Next could be ByteDance, the aggressive maker of AI-powered news app Toutiao and short-video app TikTok. Didi may take longer (see more on Didi in chapter 7 on ride-hailing services). These new dragons are breathing fire but face substantial risks. Like many hypergrowth companies anywhere in the world, steady profitability eludes them. China's BAT threesome Baidu, Alibaba, and Tencent could gobble them up by jumping into their market sectors with bigger piles of cash. Newcomers threaten to displace them if they don't keep coming up with more new thrills. How has this new generation of Chinese tech startups risen so fast, so far?

Shedding a Copycat Image

The Chinese smartphone startup Xiaomi has come out of nowhere over the past nine years to stack up $17 billion in revenues. While Samsung and Apple are the hottest smartphone makers in the United States, in Asia, Xiaomi offers strong competition to these leaders with cool-looking Android phones, which borrow some design elements from the iPhone but sell for less than half the price. Its handsets have won multiple design awards, and Xiaomi holds 7,000 patents internationally. Xiaomi has revolutionized the "made in China" stereotype by innovating fast and scaling up with inexpensive but full-featured, high-quality phones. Once dubbed the Apple of China, Xiaomi has won over many skeptics with several innovations, such as foldable phones, superthin models, extra-large screen displays, and all-ceramic phone casings.

But Xiaomi still can't shake that copycat image. Its Mi8 phone, cleverly launched in 2018 on its eighth anniversary, was labeled the "most brazen iPhone copycat yet"[1] for a near-identical look to the showstopping, $1,000-priced iPhone X and other similarities, such as a facial ID feature.

Xiaomi—A Bit like Amazon and Google?

Xiaomi is often likened to Apple, but founder Lei Jun prefers to refer to his startup as "a bit like Amazon with some elements of Google."[2] Xiaomi sells smart-connected devices online like Amazon with Alexa voice-activated gadgets and Google smart-home speakers and lights. Xiaomi phones run on MIUI or "me, you, I," a tweaked version of the Google Android operating system. Xiaomi could best be described as an Apple-plus—and even more so now that Apple is focusing on an expanded line of entertainment services. Xiaomi operates in three key business sectors. Its smartphones preloaded with dozens of music, video, and gaming apps, are well known in Asia, used by 190 million monthly internet service users. What's less known—and probably not much at all in the United States—is that Xiaomi also makes and sells a wide range of internet-connected devices, like laptops, TVs, speakers, routers, rice cookers, vacuum cleaners, fans, and air purifiers. Additionally, Xiaomi runs e-commerce site Mi.com, and operates Mi retail stores that carry its array of household and lifestyle goods in Asia and Europe. This wide network of merchandise and services sold online and offline is not so easily replicated. See table 3-1.

Table 3-1

At a Glance: Xiaomi

Location: Beijing

Founder: Chinese serial entrepreneur and angel investor Lei Jun

Launch year: 2010

Business: smartphones, internet-connected gadgets, mobile apps

Financials: $17 billion in revenues in 2018, 53% revenue growth, unprofitable

Status: publicly traded on HKSE, IPO in mid-2018 raised $4.72 billion at a valuation of $54 billion

Notable: Xiaomi is the world's fourth-largest smartphone maker; has been considered the Apple of China

---

Xiaomi's large stable of customers, dubbed "Mi fans," provide regular feedback about the latest features in online forums and communities. They are passionate about the company's affordably priced, high-quality phones, which sport innovative technologies such as AI-powered dual cameras, wireless charging, facial unlocks, a curved ceramic casing, and customized features for users. And they love the price tag: Xiaomi phones are priced affordably from $115 for entry models to $430 for a premium line—far lower than the average selling price of $800 for Apple iPhones. Avid Mi fans queue up for hours outside new flagship Xiaomi stores on opening day. At its recently opened store in London, Xiaomi staged a promotional giveaway of its new high-performance Mi8Pro phone, which is equipped with a transparent rear cover where tiny inscriptions show up offering "innovation for everyone." That may be true: Xiaomi technology for audio, which replaces speakers with vibrations on the phone, was recently adapted and refined by Chinese smartphone maker Meizu Zero for launch of its concept phones of the future (no buttons, ports, or speakers) at the 2019 Mobile World Congress.

A Bumpy Ride

But Xiaomi's short history has been bumpy. Xiaomi was flying high in going public on July 9, 2018, and raising $4.72 billion at a valuation of $54 billion—the largest in Chinese tech since Alibaba. The IPO raised far less than the anticipated $100 billion, however, during a time of escalating trade and tech tensions. An additional factor among investors was how Xiaomi should be judged—as a smartphone maker or an expanding group of online services. Months after the IPO, the stock price was hovering well below its first day of trading as global smartphone demand slowed. Profitability has been elusive, although Xiaomi posted a $1.2 billion net profit in its first quarter after the IPO.

What can't be denied is Xiaomi's impressive growth overall. Xiaomi ranks fourth among global competitors, right after Samsung, Apple, and Huawei with a market share of 8.4 percent.[3] In China, Xiaomi briefly led the smartphone market—the world's largest with nearly one-quarter of global sales.[4] But within two years, Xiaomi couldn't keep up with the superfast expansion and slipped to fourth place behind Huawei, and two local Chinese upstarts.

The core strength of Xiaomi has always been smartphones in China. Xiaomi derives nearly two-thirds of its business from smartphones and nearly two-thirds of the company's revenues come from China. But going outside China is clearly a goal, as it is with many of the Chinese tech titans. Xiaomi succeeded in winning the India market within three and a half years of entry and is expanding its presence in key European cities.

Getting into the United States remains a key goal. Founder Lei has indicated that he's been considering breaking into the US market, and he's set 2019 as the time frame. But the US market remains a distant challenge for Xiaomi. Its phones are stocked with apps suited to the Chinese market. The timing is not good at this point for Xiaomi to launch its phones in the States due to US security concerns about buying phones from Chinese makers. As an example of the frictions, US retailer Best Buy recently yanked Chinese Huawei phones from store shelves. As was the case with other Chinese tech titans, Xiaomi has turned to its more familiar home turf and has begun manufacturing its own chips to reduce reliance on US suppliers such as Qualcomm.

In 2018, I had an opportunity to meet with Donovan Sung,

Xiaomi director of product management, who with his good looks and poised speaking style is a great spokesperson in the West for the company. He was in New York City seeking to raise the company's US profile and tout its progress. From a showroom that Xiaomi's public relations team set up in midtown Manhattan, he introduced me to an array of smart-home products and gadgets, including IoT-connected headphones, cameras, speakers, thermostats, and portable power banks. Some of these products are for sale on Amazon, but it can be difficult to find Xiaomi handsets in the United States unless you look on eBay or Craigslist.

Customers who really want Xiaomi-made phones and these accessories and items can shop at its Mi Home stores in France, Spain, Italy, and London. The new European stores add to Xiaomi's more than 300 Mi Home store locations in China and several in India, plus plans to open hundreds more globally by the end of 2019.

What Makes Lei Jun Tick?

The dramatic rise of Xiaomi traces back to founder Lei's proven track record as a serial tech entrepreneur and angel investor at the forefront of China's software and internet markets. Born in a small town in Hubei province, Lei Jun, 49, was computer obsessed at a young age. He earned a computer science degree in just two years at China's well-regarded Wuhan University, and spent the early part of his career as an engineer at Chinese software maker Kingsoft in Beijing, rising to CEO within six years of joining in 1992. His dream was to make Kingsoft into a world-class tech company akin to Microsoft. Kingsoft became China's most-used office software product, but stiff foreign competition and rampant domestic piracy almost caused Kingsoft to go bankrupt. The workaholic Lei, who nevertheless has a boyish look (he sports a long side-swept parting), helped Kingsoft diversify from word processing into games and security software and managed to take Kingsoft public on the Hong Kong

Stock Exchange in 2007. After the listing Lei left Kingsoft, but in 2011 he returned as chairman, in effect channeling Steve Jobs' own return to Apple, to steer Kingsoft into the mobile internet.

Lei, whose net worth is today pegged at $9.9 billion by *Forbes*,[5] turned to high-impact angel investing and set up his own fund, Shunwei Capital, with $2 billion in assets. His circle of connections got him access to several promising startups that became winners in China's burgeoning internet market. He played a pivotal role in developing and funding online bookstore and e-commerce site Joyo.com, which Amazon acquired for $75 million in 2014. He helped to ramp up online clothing retailer VANCL with $115 million in investments. His $603 million co-investment with friends in mobile internet browser UCWeb led to an acquisition by Alibaba in 2014 at a valuation of $3.8 billion. His $1 million bet on social gaming portal YY earned him a $129 million stake following YY's Nasdaq IPO in 2012.

But Lei's biggest hit has been in getting his entrepreneurial hands dirty with Xiaomi. His vision was to make and sell a well-designed, low-priced phone that would ride on China's emerging mobile internet (thus the "mi" in Xiaomi's name) market. In 2010, he cofounded Xiaomi (which translates as "little rice" or the makings of porridge in Mandarin) with Lin Bin, a former Microsoft and Google engineer who is now Xiaomi's president, and a team of six cofounders who are either trained engineers or designers. Their formula: affordably priced phones to drive adoption, razor-thin profit margins, and continual updates based on customer and developer feedback. They didn't spend tons on marketing and advertising, relying instead on flash sales, word-of-mouth endorsements, and direct sales to consumers in limited quantities—nothing like the iconic and flashy TV commercials for Apple.

Little Rice Goes a Long Way

Within two years of its start, Xiaomi's annual sales exceeded $1 billion in 2012. By 2014, Xiaomi's sales soared to $10 billion, and it overtook Samsung, Apple, and Huawei to become the leading smartphone brand in China. But within two years, sales of Xiaomi phones dipped and its rank slipped to fourth place in China. The problem stemmed from lack of retail distribution and supply-chain shortages.

China telecom giant Huawei took the lead while two lower-cost Chinese brands that prioritized rural China sales, OPPO and Vivo, raced ahead of Xiaomi.[6]

A micromanager much like his hero Jobs and a tireless worker who puts in 100-hour workweeks, Lei led his team to a remarkable comeback in 2017. His turnaround strategy: Xiaomi invested heavily in expanding its Mi Home retail stores to more than 331 outlets across 51 China cities within five quarters, added retail to India to integrate with previous online-only channels, and broadened the distribution network to third parties. He also cultivated a Mi fan base online with community forums. The turnaround was cinched with the introduction of a popular series of high-quality, ceramic-encased Mi Mix phones with full-screen displays and a super-thin frame. Lei claims there's never been another smartphone maker that successfully rebounded after a sales decline. He and his team put in the hours—"007," a reference to all hours every day of the week—to make it happen.

Xiaomi Advantage: Hardware and Software

Xiaomi's business model is another aspect of its creativity—it's described by founder Jun as a "triathlon business model" comprised of three synergistic pillars of growth. Handset sales account for the bulk, or about 70 percent, of Xiaomi revenues; IoT gadgets and consumer goods (even spinning wheel suitcases) bring in 22 percent; and internet value-added services, such as games, account for 9 percent.

At first glance, Xiaomi may seem like a hardware company only with smartphones and smart TVs, but it's actually succeeded as the "first internet-of-things company with an array of smart hardware products," notes tech and media analyst Ben Thompson, founder of *Stratechery*.[7] Thompson points out that Xiaomi is the rare company that has succeeded in both hardware and software, adding that Alibaba and Amazon (with Kindle) have dabbled in hardware but not as a core business.

Xiaomi effectively uses the razor and blade marketing scheme of selling one item at a low cost to increase sales of a complimentary item. Xiaomi keeps the cost of its smartphones and smart household goods at a level that limits profit margins to only 5 percent. This helps to build a customer base. Then it hooks those users on its multiple apps for music, videos and games that are monetized with advertising, subscriptions, and virtual gifts. Take a lesson, Apple!

From Rice Cookers to Electric Scooters

Another twist in the Xiaomi business model comes from some 100 partner companies that it incubates or invests in. These partners are a pillar in Xiaomi's growth. They make internet-connected devices and household goods such as motion-activated night-lights and water purifiers, primed for China's rising middle class, who are outfitting their homes for the first time. Frequent product introductions and tweaks keep consumers coming back for more. Two of those partners—fitness tracker Huami (which was outselling Fitbit and Apple Watch in late 2018) and smart-home product maker Viomi—went public in New York in 2018.

90 Minutes to Push Invest Button

Xiaomi's multipronged and interconnected business model is distinct and could be considered leading edge. "It's very difficult for any investor to fully understand Xiaomi's model, which is uniquely

Xiaomi's," says Hans Tung, a board observer and an early venture investor in Xiaomi dating back to its beginning in January 2010.

As Xiaomi took off in 2014 and drew accusations of copycatting, Tung strongly disputed the idea that Xiaomi is a copy of the iPhone in interviews with me. He pointed out three key differentiators to me: Xiaomi has customizable features, it relies on social media feedback to tweak features on a daily basis, and the company originally relied on direct sales online and word-of-mouth advertising.

He also recounted what led him to invest in Xiaomi when it seemed like a crazy idea at the time. Xiaomi was a small company of only 10 to 20 employees with no hardware industry experience going up against several well-established brands. Venture capitalist Richard Liu of Hong Kong–based Morningside Venture Capital, backed by billionaire Ronnie Chan, also pulled the investment lever for Xiaomi nearly from day one. But it was the vision that founder Lei outlined, without a PowerPoint, that won over Tung in about 90 minutes:

- in the next 10 years smartphones will replace laptops
- localized and customized features will be built into smartphones that can be updated regularly
- direct-to-consumer sales channels will bypass the middleman so cost savings can be passed on
- a world-class team of overseas returnees and locals will manage the startup

In 2012, Xiaomi raised $216 million in venture capital that valued the startup at $4 billion from investors that included the existing backers plus IDG, Temasek, and DST Global founder Yuri Milner. Within four years after those initial investments, Xiaomi went on to raise $1.1 billion in 2014 from Jack Ma's Yunfeng Capital, DST Global, and others in a deal that tipped the scales for Xiaomi as the world's highest-valued unicorn at $45 billion.

In a note to employees the day before the IPO, Lei, who owns about 30 percent of the company, proudly pointed out that Xiaomi's earliest venture capital investment of $5 million has now earned a return of 866 times! "No one expected that this unremarkable, small company would go on to have such an epic and exciting entrepreneurial journey," he wrote.[8] Not only that but three of the early cofounder team of eight engineers and designers became billionaires after the IPO. Two of them had tech careers at Microsoft or Google, while the third worked with Lei at Kingsoft.

Going Outside China

Xiaomi has scaled fast by riding on China's explosive adoption of smartphones and by expanding to emerging markets such as India, where its low prices have helped it to become the top-selling smartphone in the world's second-largest smartphone market after China. International sales have risen from almost nothing a few years ago to about one-third of sales from 74 countries and regions. Xiaomi won over Indian consumers with reasonably priced phones that offer customized local features such as a heat-sensitive control to decrease the phone temperature and chargers that adjust to power fluctuations.

Hugo Barra, a Brazilian tech whiz and entrepreneur Lei poached from Google's Android team in 2013 in Silicon Valley headed up Xiaomi's international push and became the face of the company for several years. This proved to be a bonus since founder Lei is learning to speak English. When Lei traveled to India to introduce a new phone in 2015, for instance, his mangled English and awkwardly phrased opening line—"Are you OK?"—went viral on video and was widely spoofed.

Barra decided to pack his bags in 2017 and return to more familiar turf in Silicon Valley, to lead Facebook's efforts in virtual reality with Oculus and more recently, global partnerships for AR and VR. Filling his shoes at Xiaomi is Wang Xiang, a longtime Qualcomm top executive in China.

The next outward push for Xiaomi is to open 100 Mi Home stores in India before 2020. Elsewhere, Xiaomi is playing catch-up to Samsung's lead in the large Indonesian market and has become among the top five smartphone brands in Russia, Greece, Egypt, Poland, Bulgaria, Czech Republic, and Kazakhstan as well as several markets in Asia.

What's next for Xiaomi? It's getting into fintech. A new subsidiary, Xiaomi Finance, under the leadership of Xiaomi engineering cofounder Hong Feng is leveraging the company's data to offer microloans, money transfers, bill payments, internet banking, a money market fund, and financial services for small companies in its supply chain. Xiaomi is preinstalling these financial services into its smartphones. It's also investing in Indian lending startups. These moves plant Xiaomi within the turf of Alibaba and Tencent. Whether it's too much of a stretch to compete in a brand-new field and against the original tech titans is a valid question.

But it seems clear that getting into the United States, where it could compete directly with Apple, does remain a stretch, particularly given the current tech and trade frictions. For how long remains to be seen, and Xiaomi will seek growth elsewhere.

TOUTIAO: The Next News and Video King

The award winning documentary *People's Republic of Desire* captures the weird and wacky world of livestreaming celebrities in China and their search for fame and fortune online in an isolated, increasingly lonely society. Two young livestreaming stars pull in as much as $40,000 monthly with singing, dancing, comedy, and acting performances set to tracks of hip-hop, pop, rock, and electronic music and edited-in emoji stickers and animated special effects. Online fans like, comment on, and tip them with virtual gifts, such as roses that are real money. The company featured in the documentary, YY, is one of the first Chinese livestreaming platforms. I was at YY's IPO at Nasdaq in 2012, when the social entertainment service raised $83 million. I watched as founder and CEO David Li and his venture backers Lei Jun of Xiaomi, Jenny Lee of GGV Capital, and Richard Liu of Morningside Venture Capital, all proudly held up their stuffed raccoon–looking YY mascots in the middle of Times Square and cheered. Now a multibillion-dollar company, YY has made its mark in China tech. Livestreaming has become a $5 billion business, and nearly half of China's internet users have watched a livestream. YY helped to invent the livestreaming business model of making money by taking a chunk of revenue from promotional agencies and from fans virtual gifts to performers.

BYTEDANCE: The New "B"

In today's fast-moving and intense digital markets, a new, shorter form of livestreaming—15-second music video selfies—has popped up. It's popularized by Chinese upstart ByteDance, the most valuable startup in the world and a new challenger to Netflix, YouTube, Snapchat, and Tencent, as well as YY.

ByteDance founder and serial entrepreneur Zhang Yiming has a knack for anticipating content trends and leveraging artificial intelligence to bring news and entertainment to an entirely new level. His apps leverage machine learning to find out what viewers and readers prefer and personalizes a feed or stream to them that gets more precise with each use. His video and news apps have a following globally, which means that his content platform startup named ByteDance could be China's first global internet success story.

It was back in 2012, when Zhang cranked up his flagship product, a news app called Toutiao or Today's Headlines, realizing how fast newspaper readership was being replaced by digital media. A few years later, in 2016, he launched his second media app, which features 15-second video clips as Douyin in China and TikTok in the West. These user-created videos are loaded with boyish pranks, dancing lessons, dog grooming tips, and lip-synching and have caught on among teens and Millennials with little translation required. Both the news and video apps have become mainstream successes internationally, showing that Chinese business models in the consumer mobile internet can be relevant outside China.

The young creator of these wildly popular apps is delivering on his mission to, as it's worded on the startup's website, "combine the power of AI with the growth of mobile internet to revolutionize the way people consume and receive information." He claims that ByteDance is "one of the first companies to launch mobile-first products powered by machine learning technology."[9]

Both of his apps leverage AI to match up users with advertising and content. As with YY, monetization also comes from virtual gifts and a new avenue of mini-shops where merchandise can be bought directly from the app. ByteDance has become a star in the social media world and could be the next B after Baidu. It's even been ripped off by Facebook and, yes, has been accused of spreading fake news and vulgar content.

These challenges aside, ByteDance represents a breakthrough for Chinese internet services and their growing presence outside China. Its news app, Toutiao, touts 120 million daily readers. In China, where Facebook is blocked, users spend well more than one hour daily on the app, more than the average user of Facebook or Tencent's WeChat and Weibo. There's also an English version, TopBuzz, with 36 million monthly users. Its short-video platform, TikTok, has surpassed 500 million monthly active users globally.[10] And TikTok is ranked as one of the world's top downloaded iPhone apps, in the top 20 league with YouTube, Instagram, Snapchat, and Messenger.[11] TikTok got a boost internationally when its parent company bought and then merged in Musical.ly, a Chinese social video app with a large following outside China.

ByteDance founder Zhang, 36, grew up in the southern province of Fujian and graduated as a software engineer from Nankai University in Tianjin. He's used his engineering smarts, ambition, and consumer instincts to pave the way to billionaire status, ranking twenty-fifth on the *Forbes* China Rich List.[12] Fiercely independent, he's reportedly turned down an acquisition offer for ByteDance from at least one of the internet giants. He has a bigger global vision than working for Tencent or Alibaba. And, in fact, his startup beat Tencent to the short-video phenomenon.

Zhang has been down the entrepreneur road before. After working briefly at Microsoft, he started travel-booking site Kuxun, which was sold to TripAdvisor in 2009 and later resold to Chinese startup Meituan. Zhang also developed a train ticket booking program to check availability and buy tickets with short-text messages during peak travel seasons.

When he first shopped the idea of a news aggregator app powered by artificial intelligence, he couldn't convince many investors that he could outsmart Tencent or Baidu. But in 2012 he managed to persuade SIG Asia and its forward-thinking managing director Joan Wang to sign on. As the app gained traction, Sequoia Capital China next led a funding round of $100 million in mid-2014, and in April 2017 Sequoia put in $1 billion along with CCB International, an investment unit of China Construction Bank, at a $30 billion valuation. ByteDance jumped to the top of the unicorn list of most valuable startups (past Uber) in September 2018, when SoftBank and private equity firm KKR invested $3 billion at a valuation past $75 billion. The startup's investors also include General Atlantic, Hillhouse Capital, and Russian billionaire Yuri Milner. See table 3-2.

What could be next for ByteDance is going public. A comparable Chinese video streaming site themed in comics, animation, and games, Bilibili, already went public, on Nasdaq, in 2018.

Table 3-2

At a Glance: ByteDance

Founder: Chinese serial entrepreneur Zhang Yiming

Launched: 2012

Location: Beijing

Main innovation: AI-powered apps TikTok for video and Toutiao for news

Status: privately held at a $75 billion valuation, top unicorn in the world

Notable: could be China's first global internet success story

---

Zhang is riding high with the success of TikTok. It's actually similar to the US short-video–sharing app Vine that Twitter bought in 2012 and shut down four years after it failed to keep pace in the United States. You can bet now that Twitter wishes it had held on longer.

The popularity of the China-made video app has put YouTube, Facebook, Snapchat, Tencent Video, and Baidu's iQiyi on notice. Getting in on the act, Facebook has launched its own short-format video app, Lasso, which is widely considered a knockoff of TikTok. Lasso lets US users sign on through Facebook. Facebook-owned Instagram also has jumped in, with an Instagram Stories feature of 15-second and quick-disappearing videos it launched in 2016 that quickly shot up to 400 million users. Tencent and Baidu have each launched their own short-video apps in China, and Tencent has funded TikTok's key Chinese rival, Kuaishou, along with Baidu Capital, DCM Ventures, Morningside Venture Capital, and Sequoia Capital China.

If you think that's a lot to take in, then realize that the stakes are very high in this new market. China's superheated, short-video app market is forecast to reach $14.1 billion by 2020,[13] a sizable chunk of the $17.6 billion video streaming market in China.[14]

US film and TV studio executives are paying attention. *The Tonight Show* host Jimmy Fallon got in on the short-video craze when the late-night TV comedian challenged viewers to roll around on the ground to a quick clip of an old western movie. Entertainment executive Jeffrey Katzenberg is ramping up a mobile video platform, NewTV, named Quibi, with Alibaba as one of the initial investors in a $1 billion recent fund-raising effort. Katzenberg has hired former eBay and Hewlett-Packard CEO Meg Whitman to run Quibi.

Nasty Mobile Streaming Fight

In China, Tencent and ByteDance are duking it out for the market lead.

The conflicts started in 2017 when Tencent invested $350 million at a valuation of $3 billion in ByteDance lead rival, Chinese video streamer Kuaishou, started by CEO and former Google and Baidu programmer Su Hua.[15] The next year, ByteDance and Tencent founders got into a spat on WeChat Moments news stream. Zhang accused Tencent of plagiarism after the titan revived and released a video clip feature and news reading function. Unfair competition and defamation claims shot back and forth between the two. In early 2019, ByteDance moved into Tencent turf by debuting video messaging app Duoshan, which lets users share videos that disappear within 72 hours.

Baidu came out swinging too. The tech titan entered the race with short-video app Haokan and accused news feed WeChat Moments of blocking Baidu's video content.

World's First AI News Anchor

ByteDance is shaking things up in other ways, chiefly in traditional news gathering and distribution. Its Toutiao or "Today's Headlines" app pushes news and commentaries to online readers based on machine-learning algorithms of their preferences. Forget human editors. Toutiao curates personalized news from pinpointed interests and likes.

China is ahead of the United States in experimenting with AI news curation. The world's first AI news anchor imitating real-life broadcasters went live in 2018 on China's state-run Xinhua News—and appeared afterward as a gag on CNBC's *Squawk Box*. Toutiao's AI bot Xiaomingbot automatically generated news articles using machine learning during the 2016 Olympic Games. It's no wonder that tech futurist Michael Spencer contends that the content discovery and creation playbook from ByteDance is what comes after Facebook.[16]

The result of these AI content engines, however, can be clickbait-y sensational content that gets the most clicks. Childish boys staging fart competitions or a young girl eating a worm are not beneath these user-created videos, which are popular among rural Chinese residents who want personalized content well outside of stodgy state-controlled media outlets. "The content tends to lean more toward lowbrow. It doesn't appeal to folks in coastal cities, where we have an explosion of information," explained Jenny Lee, a partner at GGV Capital. "But for a farmer out in the field, or the taxi drivers, they might never leave their town for their entire life."[17]

Venture partner Connie Chan at Andreessen Horowitz in San Francisco points out the AI-powered apps at ByteDance go to an extreme not common yet in the West. TikTok uses the app's algorithms to decide which videos to show users, dictates your feed entirely, and learns your preferences the more you use it. This is different from Facebook, Netflix, Spotify, and YouTube, which use

AI to recommend posts, she notes. It remains to be seen which approach will maximize engagement,[18] observes Chan, who follows China social media trends for the venture firm.

Certainly, Toutiao is outpacing traditional news portals in volume. Its content technology sorts and tags more than 200,000 articles and videos daily and personalizes news feeds based on analysis of data obtained through the users' locations, phone model, and click history. Users open the app and access news through Toutiao's 4,000 media partnerships without following other accounts, unlike Facebook or Twitter.

Anu Hariharan, a partner with Y Combinator's Continuity Fund in San Francisco, likens Toutiao to YouTube and technology news aggregator Techmeme in one. She finds the most interesting thing about Toutiao to be how it uses machine- and deep-learning algorithms to serve up personalized, high-quality content without any user inputs, social graphs, or product purchase history to rely on.[19]

From Sea to Shining Sea

ByteDance has been moving up in recent years with content deals and smart acquisitions, fulfilling founder Zhang's mission of making his startup borderless. That goal post got a lot closer when, in November 2017, ByteDance paid about $900 million to acquire Musical.ly, a social video app based in Shanghai with more than 200 million users worldwide. The deal combined TikTok's AI-fed streams and monetization track record with Musical.ly's product innovation and grasp of users' needs and tastes in the West. The result was a multicultural DNA. After ByteDance folded the four-year-old Musical.ly into TikTok and rebranded it to a single application under the TikTok name, the app immediately gained some 30 million new users within three months. ByteDance also got inroads into Hollywood with Musical.ly and its deals with Viacom and NBCUniversal for short-form video shows.

ByteDance founder Zhang emphasized that this deal makes a lot of sense because it integrates Musical.ly's global reach with the massive user base of ByteDance in China and key Asian markets, and it creates a global digital media platform for content creators and brands both inside and out of China.

Fishing for more, the aggressive ByteDance also has been busy buying into innovators and making deals from a Los Angeles–based office that is on the lookout for more. There were 62 job openings at ByteDance in early 2019, ranging from business, strategy, and communications to engineering and product development. Over the past few years, ByteDance has snapped up Los Angeles–based Flipagram, a video and photo creation app set to music clips, and has invested $50 million in Live.me, a livestreaming app that is majority owned by Chinese mobile app developer Cheetah Mobile. Additionally, ByteDance acquired News Republic, a global mobile news aggregation service based in France, from Cheetah Mobile for $86.6 million. ByteDance even tried to buy a major stake in US social news aggregator Reddit from Si Newhouse's Advance Publications but lost that deal to Tencent, which swept in with a $300 million co-investment in early 2019.

Stretching from its news base, Toutiao is seeking to further maximize impact with partnerships and moves into e-commerce. A deal was made with US internet media outlet BuzzFeed to share its entertainment content in China. ByteDance also reached a strategic deal with China's e-commerce giant JD.com to let Toutiao users shop the e-commerce site. More than that, Toutiao is leveraging its massive traffic to take on Alibaba and JD.com in e-commerce. Toutiao recently launched an e-commerce app, Zhidian, to sell consumer and household goods.

King of Titillating Content

Like many digital media companies worldwide, ByteDance faces challenges dealing with fake news and offensive content. Chinese regulators have targeted vulgar content and "useless information"—crude comedy, scandals, and celebrity gossip, for example. That sort of content earned Toutiao founder Zhang the title of the "King of Titillating Content," as reported by the *South China Morning Post*.

Taking a proactive stance, Zhang has pledged that ByteDance will increase its team of censors, create a backlist of banned users, and improve its technology to screen content. The company has recently hired 2,000 content reviewers and banned more than 30,000 accounts on its popular Douyin video app to comply with government pressure to clean up China's internet.

More fixes are coming: an AI lab that ByteDance formed in 2016 is working on state-of-the-art innovations in artificial intelligence. The lab is headed by Wei-Ying Ma, previously Microsoft Research Asia's assistant managing director. His marching orders are to develop machine-learning algorithms to weed out offensive content and to pinpoint more personalized content recommendations.

ByteDance shows no signs of slowing down. It's building an empire of apps for a new generation. In doing so, ByteDance is challenging China's traditional BAT leaders and showing up internet leaders of the West.

MEITUAN DIANPING: The *M* of TMD

Chinese entrepreneur Wang Xing was once known as "the cloner." In the early days of the internet in China, he cloned Facebook, Friendster, and Twitter. None of them worked out great. Now he's hard at work on his near unpronounceable Meituan Dianping, which is innovating with a superapp for services. Tencent owns 20 percent of Meituan, which went public in Hong Kong in September 2018 and raised $4.2 billion. But the staying power of

Meituan as a leader is not a sure bet given it's losing money and faces increased competition from Alibaba-owned delivery service Ele.me. See table 3-3.

Over the past decade, Meituan has emerged as a titan by catering to China's burgeoning urban middle class who are using its all-in-one app to order takeout lunches, make restaurant reservations, book hotels, purchase movie tickets, and redeem vouchers for manicures and massages. This multifunctional app combines Yelp, Booking.com, GrubHub, Uber Eats, Kayak, Fandango, and OpenTable and even loops in a Whole Foods–type grocery store. There's no single equivalent to Meituan in the United States, where apps typically specialize in one vertical sector.

Table 3-3

At a Glance: Meituan Dianping

Founder: serial entrepreneur Wang Xing

Location: Beijing

Launch: 2010

Merger with Dianping: 2015

Status: HKSE listing raised $4.2 billion at a valuation of nearly $53 billion

Main Innovation: an all-in-one app for services and an AI-driven moped delivery system

2018 Gross Merchandise Volume: [20]
$76.9 billion, up 44%, from 6.4 billion food delivery transactions and 284 million hotel room booking nights in China

2018 Financials:
$9.7 billion in revenues, up 92%; $1.27 billion adjusted net loss

Notable: founder is known as China's internet cloner; this is his fourth Chinese startup

---

Meituan founder Wang lost his Facebook lookalike due to high cash burn. He wants to make sure his current business thrives. He's chasing what he sees as a $1 trillion opportunity—on-demand food delivery in China, which he says isn't an alternative to restaurant dining or preparing food at home, but a way of life. He may be right; consider how popular quick meal deliveries by DoorDash and Grub-Hub are in the United States.

Meituan founder and CEO Wang is convinced he will triumph because, as he points out, "No matter what happens, people still need to eat, and we provide the most convenient way for people to eat."[21]

With the Meituan app in China, you can track your order, see the restaurant location, tell where the delivery person is on the route, check arrival time, see the name and image of the delivery person, and call the courier directly.

Meituan deliveries usually take no more than 28 minutes from the time customers place an order to arrival, thanks to AI that cues the shortest routes. Distances are typically within a mile in China's densely populated cities. Couriers are paid about $1 per delivery and are in abundant supply. China's time-pressured working population is quite willing to pay for this convenience.

Bringing ready-to-eat Chinese lunch boxes to office workers is the busiest time.

From Scooters to Robots

Throughout Beijing and Shanghai, Meituan's fluorescent yellow-and-black clad couriers—nearly 600,000 of them nationwide—zoom by on mopeds to deliver ready-to-eat meals and merchandise. Meituan has cornered the on-demand delivery market in China with about a 60 percent share.

Late one morning at Meituan headquarters, I spot several couriers getting ready to take off with their lunch deliveries. The bustling northeast Beijing corporate headquarters of Meituan is filled with young, energetic employees, and it's hard to miss a large billboard promoting the company's feel-good mission statement: "Help people eat better, live better." Meituan translates as "beautiful together." A large showroom in the extensive lobby displays screen after screen documenting the company's evolution from group-buying startup in 2010 to merger with rival restaurant review and dining site Dianping to richly funded unicorn with big-name backers to publicly traded company in 2018. The showroom also showcases Meituan's innovations in autonomous delivery to improve efficiencies. Dr. Xia Huaxia, general manager of autonomous delivery department, shows me two self-driving delivery robots that Meituan is testing in Beijing to pick up orders from riders and deliver them to customers within the last 10 meters. The robots have separate sections for hot and cold foods.

He also tells me how Meituan is relying on big data and AI innovations. A patented dispatching technology analyzes big data to find the shortest routes and nearest couriers and to avoid traffic jams and accident sites. An intelligent voice assistant lets couriers receive and report orders when delivering without having to operate their mobile phones while riding. Such advances have helped Meituan to shave seven minutes off its average delivery times since 2016.

Other techie stuff Meituan has built in are security checks that identify and verify riders by QR code and an advanced electronic record management system to confirm business licenses of merchants on its platform by connecting with a government supervision database. The system also can synchronize data to track food safety and hygiene and check and analyze customer reviews by time period, location, and product category to spot any trouble.

The huge digital screens provide a good overview of the company's leadership of the food delivery service market that has sprung up in China with urbanization, technology development, widespread mobile internet usage, and increased consumer spending.

My vision goes blurry taking in all the info as I'm shown how Meituan stacks up in China. Meituan packs in more than 200 service and product categories. The app has racked up 400 million active buyers, 5.8 million merchants in 2,800 Chinese cities, 6.7 billion transactions, and 5 billion user reviews.[22]

The upside is clear. The consumer economy in China will account for about half of the country's GDP growth by 2020;[23] the e-commerce market will hit $1.8 trillion by 2022, up from $1.1 trillion in 2018.[24] Meanwhile, the food services segment of the e-commerce market is growing by nearly 20 percent yearly and could top $1.15 billion by 2023.[25]

Never Made Money

What these screens don't divulge is that Meituan has been losing money since its inception. Like many fast-growing tech companies in developing markets, the priority is on grabbing market share rather than making money.

The problem is that Meituan's core business of food delivery is labor intensive and cash-burn-heavy and operates on thin profit margins, points out Eleanor Creagh, market strategist at Saxo Capital Markets. Meituan faces intense competition from Alibaba-backed Ele.me in food delivery and can't charge more for deliveries without losing customers. The battle for market share is costly, and spending needed to attract and retain users through generous subsidies and incentives will remain high, she notes. Meituan's travel and hotel segment of the business is far more profitable, with gross margins of 88 percent, Creagh concludes.[26]

An executive at a major Chinese competitor contends that

Meituan's business unit economics and losses are unsustainable and believes Meituan will be outspent by rivals. Going public was a big mistake for Meituan, since its poor financials were open for all to see, the executive at a rival company noted.

About two-thirds of Meituan revenues are for food delivery services, while travel bookings and such services as wedding planning account for most of the remainder. To avoid capital-intensive subsidies to win customers, Meituan is going after merchants. Meituan is bringing in additional revenue from merchant commissions on customer orders and fees for online marketing, advertising, and business services such as management of payroll, inventory, customer relations, and microloans at a monthly interest rate of 1.5 percent.[27]

Meituan also has branched out into other businesses but with mixed results. The company entered the bike-sharing space with a power-sized acquisition of Mobike but downsized operations a few months later to avoid oversupply in the cooling bike rental business, improve operating efficiencies, and curtail losses. That seems smart. An expansion into ride-hailing operations in Shanghai and Nanjing to test the idea of letting restaurants pay for the rides to their locations has been cut short. That also was a smart move (read more about ride-hailing companies in chapter 7). Meanwhile, Meituan has recently started to explore getting into the gaming business—maybe not so smart, given Tencent's recent (bad) luck with this sector.

Three Times, You Win

From Fujian province in southeastern China, a part of the country known for its successful small business culture, Wang Xing, now 40 and balding, has become one of China's richest individuals, with a net worth of $5 billion.[28] Wang earned a bachelor's degree in electronic engineering degree from China's MIT, Tsinghua University, then went to the United States for a master's degree in computer science at the University of Delaware in 2005, but dropped out of PhD studies there to return home. Inspired by the success of Facebook, he embarked on a winding entrepreneurial path during the rise of the internet in China. He's faced many setbacks, but Meituan could be his victory. As Wang says in a poetic way about his journey, "The more faithful we are to the future, the more patient we are."

His first effort at building a social network, a basic copy of early social networking site Friendster named Duoduoyou, didn't catch on. His Facebook copy, Xiaonei, was popular with users but burned through cash quickly and was sold to Oak Pacific Interactive for $2 million in 2006 and renamed Renren. His next try was Fanfou, a popular Twitter copy. But Fanfou was taken offline for 18 months by Chinese government censors in 2009 during a series of violent riots in western China. Meanwhile, Fanfou was soon overtaken when Nasdaq-listed Chinese highflier Sina Corp. launched microblogging website Sina Weibo.

The entrepreneurial programmer Wang was taking a clue from Silicon Valley that it's okay to fail many times when he launched Meituan in 2010. Meituan actually began as a copy too, of US group discount-buying site Groupon. That wasn't a sure bet either.

Groupon lookalikes were springing up throughout urban China in what was labeled the Battle of a Thousand Groupons, all chasing China's bargain-loving consumers. Big-time investors Alibaba and Tencent and several Chinese venture heavyweights poured millions into the copycats, eager to cash in on the craze. The original Groupon also joined the fray in 2011 in a joint venture, Gaopeng, with Tencent in China. From 2010 to 2014, Meituan raised $270 million in several rounds from Sequoia Capital China, Alibaba, and private equity firm General Atlantic. Meanwhile, from 2010 through 2012, its lead rival, Dianping, took in as much as $200 million from Sequoia Capital China, Google, Qiming Venture, and Lightspeed Venture Partners. Then in 2014, Tencent upped the ante by buying a 20 percent stake in Dianping. Also in the running was Lashou, backed by GSR Ventures and Norwest Venture Partners. Another contender, Nuomi, was ramped up by Renren.

In the battle of 1,000 Groupons, lines were drawn over larger and larger advertising budgets, bigger and bigger subsidies, and higher-value coupons offering discounts of as much as 60 percent. The lead contenders, Meituan and Dianping, kept raising capital to do battle. Meanwhile, the costly marketing expenses took a toll on the weakened. Lashou failed to go public and was bought by Chinese retail holding company Sanpower Group in 2014. Renren's daily deals site Nuomi was sold to Baidu in 2014 and faces declining market share. As for the original Groupon China, it went on a hiring and marketing spree in China and did a joint venture with Tencent but flopped—a déjà vu of earlier attempts by eBay, Yahoo!, and Google in China to win against fierce local rivals. Meanwhile, the two survivors, Meituan and Dianping, continued to fight it out with bigger and bigger pockets: Meituan drew in $700 million in 2015 and Dianping $850 million the same year.

A Golden Week Merger

It was China's Golden Week in October 2015 when a truce was called. Meituan merged with its top rival Dianping in a $15 billion transaction supported by their big-name backers Alibaba, Tencent, and Sequoia Capital China. The merger was a good fit—a mash-up of Meituan's Groupon-type vouchers for movie tickets and travel bookings and its food delivery service with the restaurant reviews and listings of Yelp-like Dianping. The combination created China's dominant group-buying leader and settled the question of whether there could only be one winner in this sector (the answer is yes). Wang took charge of the merged entity, looping in several of its prior cofounders from the internet era. He didn't waste time. Meituan Dianping in 2017 raised $4 billion in a round led by Tencent and the Priceline

Group (renamed Booking Holdings) in the United States. That financing valued the Chinese services app at $30 billion, then the world's third most valuable unicorn. Meituan emerged as China's leader in the "Internet Plus" market, or a combination of online and offline worlds known in China as simply as O2O, the hottest buzzword of its time. Now Meituan faces Alibaba's recently merged food delivery and online services Ele.me and Koubei units, which are pumped with $3 billion in funding led by SoftBank. I can't think of another series of tech deals in the United States so packed and influential.

I well remember Wang speaking at a Silicon Dragon forum in Beijing in 2012, already setting himself apart with his smarts and driven but low-key manner. He certainly pushed the supercharge button. Soon after raising that $4 billion load, in September 2018, he hit the gong to signal the company's debut on the Hong Kong Stock Exchange. Meituan raised $4.2 billion—the second-largest tech offering in Hong Kong following the $4.7 billion IPO of Chinese smartphone maker Xiaomi just two months before.

"This [Meituan] may be the most important decision in our investment journey in more than 10 years," said venture capitalist Neil Shen, founding and managing partner of Sequoia Capital China, one of the larger 12 shareholders in Meituan with about a 12 percent stake. Sequoia stands to earn nearly $5 billion on its investments of $400 million in Meituan. "In this scuffle, Wang Xing led the team to fight more and more bravely, and it was a bloody battle in the fierce competition."

Shen relates that Sequoia hasn't sold any shares in Meituan. "We see Meituan as a long-term champion in the internet space in China. I still think the growth is coming."[29]

For now, Meituan is maxed out focusing on the domestic China market. But it's starting to strategically invest in Southeast Asia, with recent buys into food delivery startup Swiggy in India and scooter market leader Go-Jek in Indonesia.

A *Little Elephant that Could*

Digitalized retailing is yet another foray for Meituan. It's opened seven so-called Little Elephant grocery stores, which resemble a mini Whole Foods—or, in China, something like its competitor's offerings: Freshippo from Alibaba and 7Fresh stores from JD.com. Meituan opened its first Little Elephant in 2017, about the same time Alibaba and JD.com launched their own retail outlets.

I visited Meituan's inaugural Little Elephant market in Beijing and was impressed with its large selection of fresh fruits, vegetables, and seafood, as well as sundry items typically stocked in a CVS or Walgreens. Little Elephant offers quick deliveries within a 3.5 kilometer radius, items tagged with bar codes, automated checkouts, and online payments, much like 7Fresh and Freshippo (but without the robots bringing food to your table). During my walk around the store, Chinese box luncheons were being prepared for takeout and dining onsite. Several young female office workers sat on stools near the open kitchen eating their lunch. I ordered lunch there too but was in a hurry and had to eat my box lunch of rice, vegetables, and chicken during the taxi ride to my next stop with another China tech titan: ride-hailing service Didi, the Chinese company that famously took over Uber in China.

CHAPTER 4

FEW US COMPANIES CRACK THE CHINA CODE

It's the rare American internet company that has succeeded behind the Great Wall of China, but Starbucks, Airbnb, WeWork, and LinkedIn keep trying harder with digitally savvy strategies borrowed from China and localized teams.

In China's large cities, Starbucks customers skip the line. Instead, they click on a smartphone app to order remotely, pay by Alipay, and get their favorite brew steaming hot for a delivery fee of about $1.30. Starbucks has gone local in China, with a digital technology and retail makeover from e-commerce giant Alibaba and its courier service Ele.com.

The US coffee chain needed a jolt when its China sales went lukewarm after a homegrown Chinese competitor, Luckin Coffee, launched in late 2017 and started eating into Starbucks' market share. Luckin took off with steep pricing discounts, on-demand mobile ordering and payments, and quick takeaways or speedy deliveries from tucked-away kiosks for on-the-go office workers. A cup of Luckin costs $3, not counting heavy subsidies such as two-for-one discounts, while Starbucks is priced at $3.50—and wins on taste.

Starbucks' new techie, digital commerce formula and partnership with Alibaba in China was cooked up by longtime friends Jack

Ma of Alibaba and Howard Schultz of Starbucks as a way to get sales percolating again. Traditional brick-and-mortar Starbucks stores weren't going to cut it among China's Millennials and their advanced mobile commerce habits.

Starbucks' sales have picked up in China since going all in on digital technologies and introducing spill-proof cups, but growth is slower than before Luckin Coffee turned up the heat in 2018. Competing against a highly caffeinated Chinese competitor is tough.

The appeal of Starbucks has always been well-designed stores where customers can linger as long as they want, sipping great-tasting coffee. Its new mobile ordering, payments, and on-demand deliveries bring it into a new era fit for China. Interestingly, Starbucks is bringing its new quick coffee service, introduced first in China, to six US cities, joining forces with UberEats food delivery service.

Starbucks' new initiative with Alibaba in China is far-reaching and introduces many firsts for the American coffee chain. A virtual store especially for China lets Starbucks customers order their favorite brew, purchase gift cards, redeem benefits, and buy mugs and coffee beans from inside Alibaba e-commerce apps with Alipay. Starbucks also has opened dedicated kitchens in Alibaba's digitalized Freshippo supermarkets in Shanghai and Hangzhou and is already in 2,000 stores in 30 Chinese cities, fulfilling large volume orders and making deliveries within 30 minutes. These mass-brewing kitchens supplement Starbucks retail sales of handcrafted coffees served by baristas.

The Seattle-based coffee chain entered the tea-drinking nation of China two decades ago and won over consumers, who developed a snobbish taste for coffee and didn't blink at paying $3.50 for an Americano, a small price for the status of a premium American brand. Now Starbucks has upgraded to China's new tech to retain its lead.

"Starbucks is growing and innovating faster in China than anywhere else in the world. Our transformational partnership with Alibaba will reshape modern retail, and represents a significant milestone in our efforts to exceed the expectations of Chinese

consumers," Starbucks president and CEO Kevin Johnson pro-
claimed in unveiling the new partnership with Alibaba.[1]

Starbucks Reserve Roastery in Shanghai Shines

I visited the Starbucks flagship Reserve Roastery on busy Nanjing
Road in Shanghai, which is one of four, including New York, Seattle, and Milan, and is the largest Starbucks outlet in the world. The
bustling two-story, 30,000-square-foot store is sort of a Disneyland
for coffee. I watched as burlap bags were coming in on a snaking
conveyor belt and coffee beans were unloaded for roasting in small
batches, stored in a huge copper cask to ensure the best flavor, then
sent through copper pipes to silos at the coffee bars or packaged
in-store for distribution to Starbucks stores in China. At a long bar,
baristas were serving up espresso, lattes, cappuccinos, and many
specialty drinks. Waiters were passing by to take orders for coffees, beer, wine, and focaccia sandwiches. Chocolates not usually
found in Starbucks were prominently displayed, enlightening Chinese consumers on what pairs well with coffee. In a nod to Chinese
tastes, there also was a well-stocked specialty tea bar. An augmented
reality platform accessed on Alibaba's app displayed information
about key features in the Reserve Roastery and a bean-to-cup story.
Customers were lingering, laughing, meeting friends, and checking
their mobile devices. The store was crowded, and a line was forming
outside.

The Shanghai Reserve Roastery opened in December 2017 and is
Starbucks' bid to keep its premium image alive in China. China has
been a sweet spot for Starbucks for many years. Starting from 1999
with one store at the China World Trade Building, Starbucks has
opened 3,700 stores in 150 Chinese cities and is on pace for 6,000
stores by 2022. China is Starbucks' fastest-growing and largest market after the United States, and represents 10 percent of the chain's
global sales.

"For nine consecutive years we lost money in China," Howard Schultz, Starbucks cofounder and former executive chairman told shareholders at a recent annual meeting. "And there were so many people who doubted whether in a tea-drinking society we could break through. Not only have we broken through, but China is going to be the largest market in the world for Starbucks."[2]

Forget Instant Coffee

While tea remains the traditional drink of China, a coffee culture has caught on in urban areas and among young professionals who enjoy hanging out at coffee shops and appreciate the finer things in life. Starbucks' green logo is easily spotted throughout many Chinese cities. The US specialty coffee shop dominates the China market with more than a 50 percent share but faces increasing competition from other international brands, Canada's Tim Hortons and the UK's Costa Coffee, new distribution outlets at supermarkets and hypermarkets, and, most of all, the rise of Chinese upstart Luckin Coffee.

The first cup of Starbucks in China was sold in 1999 when Asian private equity leader Ta-lin Hsu, founder of H&Q Asia Pacific, bought a local chain that held the license to sell Starbucks in Beijing and Tianjin, when China was opening up to economic reforms, consumerism, and American premium brands. Starbucks won favor by selling traditional Chinese moon cakes in glowing lantern boxes during the Mid-Autumn Festival. And it managed to recover from a public-relations disaster. Starbucks opened a coffeehouse that sold lattes and frappuccinos in the heart of Beijing's historic Forbidden City for nearly seven years until local protests grew so loud that the controversial outlet was closed in 2017. Showing its continuing commitment to China, that same year, Starbucks purchased the remaining 50 percent of a joint venture business in China from two Taiwanese food conglomerates for $1.3 billion, moving to full company-owned stores. Starbucks has become one of the US champions in China, and it intends to stay ahead despite coffee clashes.

Competitors are chasing big growth potential as coffee drinkers upgrade from instant coffee packets and China's $3.2 billion coffee retail market grows at a double-digit rate, projected to reach $11.5 billion in 2022.[3] Drinking coffee still isn't a daily habit in China[4]—on average, only four to five cups per year compared with one coffee per day in the United States.

Finger Luckin' Coffee

Luckin Coffee is waking up the Chinese coffee market. From opening its first store in January 2018, Luckin, as it's known, has expanded to 2,000 locations in 30 cities, aiming to surpass Starbucks. Luckin is flush with cash from raising $400 million in 2018 at a unicorn valuation from Singapore sovereign wealth fund GIC, investment banking firm China International Capital Corp., and Joy Capital. The startup recently went public, but there are a lot of questions, particularly about its cash-burning strategy to beat Starbucks.

"What we want at the moment is scale and speed," Luckin's chief marketing officer Yang Fei said at a Beijing press conference.[5] "There's no point in talking about profit."

The business model for Luckin is remarkably similar to Uber's on-demand service, except it's for coffee, not car rides. The connection is that founder Jenny Qian Zhiya is a former operations executive at on-demand chauffeured car service UCAR in China. She built that business on smart ordering and dispatching, leveraging data like Uber does to route drivers to passengers and their destinations.

It's questionable that Luckin can unseat longtime leader Starbucks in China, but it gained momentum with a recent Nasdaq IPO that raised $561 million. Starbucks is leveraging its first-mover advantage in China, global resources, and newfound tech savvy to keep an edge over Luckin. But Luckin continues the pursuit. It wouldn't be the first time that a homegrown Chinese upstart beat a major American brand in China.

Breaking Through in China

Localizing China operations, partnering, and embarking on new digital frontiers like Starbucks has is a formula that is working for a few other American tech standouts that have entered the Middle Kingdom: LinkedIn, Airbnb, WeWork, and Evernote.

Many US tech companies have stumbled in China: Uber, eBay, Yahoo!, Amazon, Groupon, and the list goes on. The most prominent fumble was Uber, which quit after a three-year, enormously costly battle by merging its operation into local Chinese rival Didi in a $35 billion deal. Underestimating dragon-powered competition has invariably been the underlying mistake of US companies pursuing increased sales from China.

The capitalistic lure of winning over hundreds of millions of Chinese consumers keeps American companies trying to get beyond the Great Wall. Censorship is the biggest obstacle for giant US internet companies that do business nearly everywhere in the world except China. Google has been exploring a reentry into China with its internally developed Project Dragonfly, which would bring a censored search engine to China, a controversial initiative that has drawn criticism from Washington DC, and some of Google's own employees, which could squelch plans. Facebook has reportedly been working on a software tool that would let the social network be censored and thus gain access to the China market.[6]

LinkedIn Deals with China Laws

Networking service LinkedIn is one of the rare internet content companies from the United States that's still accessible in China, where Facebook, Google, Twitter, and Pinterest are blocked. By focusing on professional networking, LinkedIn remains open, and you can click on it in China to share posts, messages, and content, which I've done without being blocked.

LinkedIn was attracted to China by the rapid development of the Chinese economy and growth of professionals to more than 140 million who represent roughly one in five of the world's knowledge workers. "The expansion of our offering in China marks a significant step forward in our mission to connect the world's professionals to make them more productive and successful," wrote LinkedIn's CEO Jeff Weiner in launching a simplified Chinese website in 2014.[7] "Our goal is to connect these Chinese professionals with each other and with the rest of LinkedIn's 277 million members in over 200 countries and territories."

LinkedIn China has done a lot of things right. One was its Chinese language version with localized content. Another was not going in alone. LinkedIn CEO Weiner set up a joint venture with two well-connected local investment companies, Sequoia Capital China and China Broadband Capital Partners. Experienced tech entrepreneur Derek Shen was hired to head up LinkedIn in China, bringing deep knowledge as a former VP at social networking site Renren (then China's Facebook), a founder of group buying site Nuoumi (acquired by Baidu), and former head of business development at Google China.

Shen had the autonomy to tweak the site. He made sure that local features were integrated so members could import their contacts from Weibo (China's Twitter) and link to their accounts on WeChat to share content across networks.

In less than four years, LinkedIn China grew revenue by eight times and increased membership from 4 million to 41 million users in China. That's a small fraction of the more than 500 million global users for the business-oriented social networking service that Silicon Valley entrepreneur Reid Hoffman started in 2003. But that gain counts as a big achievement in the intensely competitive Chinese marketplace, that saw two of LinkedIn's local rivals shut down.

In expanding to China, chief executive Weiner acknowledged the challenges of privacy issues and government-imposed censorship. LinkedIn promised to protect the rights and data of its members, but it had to censor content despite getting flak for doing so. Remember that a decade ago, Yahoo! cofounder Jerry Yang ran into privacy issues in China and was hammered for turning over the names of two Chinese bloggers who were later jailed.

LinkedIn has bumped up against Chinese regulations that limit online member groups and restrict individuals, though not companies, from posting job listings without verifying their identity through phone numbers. It's also dealt with requirements that make LinkedIn members with Chinese IP addresses link personal mobile numbers to their accounts to gain access to the main page. Almost all apps in China require a mobile number instead of email, which is going away fast in an era of texts and chats. Wi-fi internet access in many public places in China, including the airports, requires a local mobile phone number.

As LinkedIn China was struggling to deal with such issues and connect with and win over a fan base locally, the company faced a blow when local manager Shen resigned in mid-2017 to rejoin his former team and angel invest in a Chinese shared housing startup, Danke Apartment. Recruited to replace him was another good hire, Jian Lu, former CEO of an e-learning subsidiary of Hujiang and former CTO for video units at internet security company Qihoo 360.

But a key challenge remained—how to reach a mainstream audience, especially as competition was mounting from fast-growing, Chinese professional social networking startup Maimai, well-funded by DCM Ventures and three other investors. Additionally, Tencent and Alibaba both moved into the enterprise market with apps for office productivity, business connections, and collaborations: WeChat with Qiye Weixin, and Alibaba with DingTalk.

Notably though, LinkedIn China has survived other Chinese startups in the same sector, basically defunct copycats of LinkedIn. French social network Viadeo closed its operations in China in

2015 after acquiring Chinese startup Tianji in 2007. Another, Shanghai-based Ushi, founded by serial entrepreneur and former Hong Kong private equity investor Dominic Penaloza, almost made it. Ushi, which means "outstanding professionals," reached 200,000 members within one year of starting up in 2010, relying on word-of-mouth endorsements from an exclusive group of initial members who were invited to join and clever promotions such as a cobranded marketing program with the *Wall Street Journal China*. I joined as a member too and found Ushi a handy tool for connecting primarily with expats in China's tech and venture circles.

All looked good when Ushi snared $4.5 million through a minority investment from New York–based business connections platform Gerson Lehrman Group and then partnered to add in a Quora-like question-and-answer feature.

But as Penaloza tells it, the site was under pressure from investors to scale up and was achieving viral growth, but it couldn't retain enough users. Penaloza concludes that the China market for in-house recruiting and do-it-yourself talent searches was at an earlier stage than he had initially assumed. "It doesn't make me feel (much) better to note that none of the other 'LinkedIn of China' startups got any major traction either, including LinkedIn itself," he notes.[8] "As of 2016," he later added, "there is still no LinkedIn of China with major traction, including LinkedIn itself, which has taken a low-investment in China as it watches the market develop."

It Takes Guts to Win

What separates the winners from the losers? It comes down to several factors: understanding the Chinese culture, customizing services for Chinese consumers, giving authority for local managers to make quick decisions in the fast-paced market, partnering with China-based companies, and perhaps most of all, having the guts to compete with China's ferocious entrepreneurs. See table 4-1.

Table 4-1

Pointers to Win in China

- Find a local Chinese partner; don't go it alone.
- Hire a local team experienced in China business and tech.
- Give the local team autonomy to make decisions to build their own business model and operate independently of US headquarters.
- Customize services and features for Chinese customers.
- Strategize and then move on acquisitions that can jump-start the business.
- Learn to negotiate with demanding Chinese customers and don't expect to win every time.
- Aim for high growth before profits in China's supercharged markets.
- Develop out-of-the-box, fun promotional strategies that can stand out in a cluttered environment.
- Be prepared for sudden rule changes in China.
- Keep the long-term vision and perspective; don't expect to go public tomorrow.

WeWork and Naked Hub

New York–based WeWork arrived in China in 2016 and jumped right into China's booming office-share market, forming a stand-alone entity, WeWork China. What has worked exceedingly well for WeWork China is well-plotted acquisitions and steady, calculated expansion supported by megafunding. Within two years of plunging into the foreign market, WeWork bought Chinese coworking startup Naked Hub from Shanghai-based luxury resort operator Naked Group for a cool $400 million, a move that gave WeWork an immediate lift. WeWork scooped up 25 Naked Hub locations in

Beijing, Shanghai, and Hong Kong to add to its own 13 spots in China.

It didn't take long for WeWork CEO Adam Neumann to take a lesson from Naked Hub's operation in trendy Xintiandi, which was making bundles of money from renting office space on a flexible, come-as-you-like basis, with no monthly contract. WeWork quickly launched the feature as WeWork Go, which lets China-based customers check for open desks by mobile app, go to the location, and register by QR code. From there, the clock starts ticking—15 yuan per hour ($2.50) and double that rate for premium locations. WeWork China reportedly picked up 50,000 registered users in Shanghai after a three-month pilot of the hourly rate that compares with a monthly rental charge of $270. Founder Neumann is rolling out this pay-as-you-go feature in the United States, initially in Manhattan.

WeWork has rebranded Naked Hubs in China, but it may be a while before the funky hip decor that's popular with the locals is redone into WeWork's trademark industrial black and white glass design. A Naked Hub location that I checked out in central Beijing even sports a swimming pool.

My group, Silicon Dragon, has held tech venture events in both WeWork and Naked Hub locations in Shanghai—no complaints over the spacious and upbeat spaces but there were some surprises. A flagship location of WeWork at Weihai Lu, where we staged a program, is a former opium factory totally transformed into a stunning light-filled atrium surrounded by several levels of floors with meeting rooms and desks.

China's coworking market has taken off with the entrepreneurial boom—part of the sharing-economy fad of shared bicycles, umbrellas, basketballs, living spaces, you name it. These state-of-the-art coworking locations with hot desks, conference rooms, and kitchens are packed in Beijing, Shanghai, and Hong Kong. The race is on to bulk up and bring in more founders, startups, and freelancers.

A shakeout is looming in an increasingly crowded shared-office sector, which has disrupted the traditional workplace. Price wars have erupted. As many as 40 Chinese coworking companies that were expanding recklessly and didn't control costs have recently vanished. One of the larger operators, Kr Space in Beijing, owned by China tech news site 36Kr, was racing to beat WeWork but has recently cut staff and scaled back ambitious expansion plans.

Fresh funding is the difference between survival and closure. Deep-pocketed WeWork and its China subsidiary are loaded with new capital: WeWork China pulled in $500 million in 2018 led by SoftBank, a huge $4.4 billion investor in WeWork itself, on top of $500 million the year before. WeWork has covered China with 60 locations across a dozen megacities, adding to its global presence in 23 countries worldwide. WeWork's recent rebranding as We Company signifies its new direction into social lifestyles and even residential rentals, an elementary school, and a coding academy, and this concept could be brought to China.

But WeWork has its work cut out for it in China battling interlopers with big visions. WeWork sued a Chinese competitor, Beijing-based coworking outfit URWork, for trademark infringement, in a case settled in 2018 when URWork rebranded as Ucommune. The founder of Ucommune, real estate veteran Mao Daqing, is known as the king of China coworking. Starting from 2015 and backed by $650 million in funding from majors Sequoia Capital, ZhenFund, Matrix Partners China, and Sinovation Ventures, Ucommune has opened 120 locations in China and claims to be the nation's largest office-sharing provider. It's also tech savvy: access to one Ucommune location I toured in Beijing is by facial recognition.

Ucommune has been powering up with a string of acquisitions, grabbing smaller Chinese coworking players Wedo, Fountown, Workingdom, New Space, and Woo Space, expanding in key Asian cities, and buying Jakarta-based Rework. It's also moving stateside, penetrating Los Angeles, Seattle, and Wall Street in a joint venture with Serendipity Labs in an office tower owned by China conglomerate Fosun.

Coworking is evolving as more than hot desks, free coffee, and networking events, moving to data-driven technologies and services that can supplement revenues from renting space with goodies. The office of the future is about smart IoT technologies to connect every desk, sensors for heat and light controls, computerized cameras that can also track occupancy levels and usage, and online community building. China is at the forefront of this development too, as with many other areas where technology has filtered in.

> *"In China, what we are doing is known as fast-response tech."*
>
> **Mao Daqing**
> Founder, Ucommune

Ucommune's founder excitedly told me about his own company's evolution from a typical coworking space into a technology platform. With its recently introduced UBazaar mobile app, which is similar to WeWork's mobile app and Services Store, members can book desks and conference rooms, access content, e-commerce, financial advisory, and ads, plus get help with legal, human resources, and tech services. At Ucommune, the truly innovative idea is its connected desks, which function as information portals with built-in facial recognition and cloud computing so members can sit down, exchange information, and access their work and communities. "We never treat our business as real estate. In China, what we are doing is known as fast-response tech," Mao told me. "It's not about the space or the square meter. You can move these intelligent technology tables everywhere, to office lobbies or airport lounges."[9] Ucommune is working with a tech team to test a conceptual space in Beijing that Mao says will integrate these technologies and can operate without any management present.

Like Starbucks, WeWork may find it tough to outmaneuver a clever Chinese challenger that is smartly integrating technology tools.

Not Home Alone: Airbnb in China

This being China, there are many jaw-dropping places to experience the country's history and heritage—and Airbnb wants to make sure that visitors discover them. Chenyu Zheng has written an entire book about her adventures staying at Airbnb accommodations off the beaten trail.[10] She's stayed at an artist's home in Guilin, where her host showed her around his studio in a deserted 1970s Soviet Union factory. She's stayed at an architect's Qing dynasty home in Yangshuo, a remote village in southwest China with a view of the mountain peaks that are printed on China's 20 RMB currency. Amazing!

At Airbnb in China, guests can learn how to make Chinese soup dumplings or listen to Chinese opera, and then share their stories of experiencing local, authentic China on a content site that Airbnb developed to create awareness and inspire travelers to China to try out and get accustomed to the home-sharing concept.

Airbnb is really making an effort in China, but those efforts aren't always successful. A promotional campaign recently offered 100 Airbnb free nights and experiences to Millennials who didn't have a chance to travel within the past year, and got 10,000 sign-ups. But another promotion featuring an essay contest for four winners to fly to China and stay in a watchtower on the Great Wall bombed. The contest, planned with a state-owned tourism agency, was abruptly halted after online objections of Airbnb's plans and government disapproval.

As Airbnb prepares to go public in New York, its China journey will become more central to its success. The San Francisco–based home-share startup entered China in 2015 with a localized operation, supported by Chinese venture capital firms Sequoia Capital China and China Broadband Capital, GGV Capital, Horizon Ventures, and Hillhouse Capital. Two years into its entry, Airbnb rebranded with a Chinese name, Aibiying, which means "welcome each other with love"—although the name was panned in China on microblogging sites for being goofy and too difficult to pronounce.

The short-term rental market is still in its early stages in China. Domestic Chinese consumers aren't accustomed to the idea of renting a room to a stranger. Airbnb's appeal is mainly among Chinese tourists who stayed in home-shares when traveling abroad. But gradually, the concept is growing as Chinese tourists return to their homeland, became hosts or guests, and introduce this Western idea to China. Airbnb hasn't veered from its global branding, which emphasizes premium quality but also personalized, adventure-filled travel.

Airbnb China's recently appointed president, Tao Peng, has the autonomy to make decisions locally, move fast, and capitalize on the opportunities. He's predicting super growth based on expectations that, by 2030, China will be the world's biggest tourist destination, replacing France, as more middle-class Asians spend more on travel and as China's 400 million Millennials get into sightseeing. He expects China will be Airbnb's number one destination market globally by 2020 and Chinese tourism its largest source of business.

Today, he oversees a 200-person team in China and some 200,000 rental homes and apartments in Chinese locations, tiny but scaling up out of Airbnb's large base of 5 million homestay listings globally. The quality of Airbnb China listings is similar to those anywhere in the world, but guests and hosts in China tend to be younger, the Millennial generation who are attuned to Western culture.

The opportunity in China is massive, since tourism is a major force in the country's growing economy.[11] The most popular destinations for Airbnb guests are Shanghai and Beijing, and Shenzhen is seeing the fastest growth.

Going local, Airbnb reviews every listing, monitors the quality of listings, inspects homes, and counsels hosts on design and in-home services. For hosts of more luxurious properties, Airbnb provides decor consultations, photography tips, and top placement in search results. Airbnb offers an academy for hosts to learn the dos and don'ts of hospitality by attending workshops, offering live chats on WeChat, and watching educational videos.

Digging deeper into the Chinese market, Airbnb recently launched a Plus service in China that features beautiful homes, gracious hosts, and hotel-like features for peace of mind. Offered in 12 China cities, the collection is inspected and verified in person, using a checklist of 100 items for cleanliness, comfort, and design. Airbnb also runs a dedicated bilingual China customer service team. It's leveraging WeChat too, investing in social customer service on the social media app and letting travelers book Airbnb stays through WeChat. Airbnb also has extended its Experiences global platform to China, where hosts are selected to offer dining, arts, and cultural adventures. In two years' time, 1,000 Experiences have been offered in China.

Following the Rules: Sharing Data

Airbnb has faced tightened regulations over home rental stays in the United States and Europe, and China isn't an exception. Doing business in China means following the rules. Airbnb shares data on guests and hosts with various Chinese government agencies and stores the details on local servers. Hotels in China also hand over guest details. Still, international tourists used to the informal nature of shared rental accommodations may find this surprising from Airbnb.

Working with the Chinese government, Airbnb has provided guidance on establishing standards for short-term rentals as a safe and positive environment. It's also partnered with the China Tourism Academy on boosting rural community tourism through home sharing, identifying and offering economic opportunities for locals to become hosts in scenic rural areas like Guilin. Such localized aspects of its Chinese business probably were never imagined by cofounder and billionaire Brian Chesky when he dreamed up the homestay concept with two cofounders in San Francisco in 2008.

Airbnb had a running start in China because of some natural advantages. It's in the travel business and is inherently a global company. Airbnb wisely hired from within the region rather than fly in newbies from California to run China operations. But management turnover in China has haunted Airbnb.

The landing team for Airbnb in China was Henek Lo from investment trading positions at JPMorgan and Macquarie Group in Hong Kong and Robert Hao, a cofounder of two Hong Kong–based e-commerce startups. Within a few years, they grew the team to more than 100 and turned China into Airbnb's fastest-growing market globally by focusing on partnerships, policy, branding, social marketing, community building, and inventory growth. They also tweaked the platform to function in China. Instead of relying on Airbnb's usual partners, Google and Facebook, which are both blocked in China, they linked Airbnb to Chinese app stores, social networks, and WeChat payments. But both Lo and Hao left the company, leveraging their China grassroots experience to form Hong Kong advisory HYPE in 2017 to help startups unlock the Asian market.

In response, Airbnb poached Hong Ge from a Facebook technical director post to lead China operations. But he left in four months after it was revealed he was having an affair with an employee.[12] Airbnb's cofounder and chief strategy officer from San Francisco, Nathan Blecharczyk, next stepped in as chairman of Airbnb China.

Blecharczyk traveled to China about once a month from Northern California to check on operations and steer direction. He started learning Mandarin but admits he's no master yet.

A search was ongoing for about a year for a lead at Airbnb China; it can take a while to find top-quality talent in China's managerial ranks. Airbnb China finally tapped Tao Peng, a serial entrepreneur experienced in the travel sector, as president in mid-2018. Tao is the founder of mobile travel guide app Breadtrip, backed by Tencent and Rupert Murdoch, and homestay hosting platform City Home, which Airbnb has funded with $5 million and could integrate. Cofounder Blecharczyk continues to travel regularly to China, and remains chairman of Airbnb China, working closely with Tao and serving as an ambassador between local and global operations.

In China, Airbnb faces two major local rivals, both backed by Chinese heavyweights and in different segments of the sector. Homegrown Tujia focuses on professionally managed serviced apartments, often flats left vacant by property developers and individual owners in China's real estate building boom. Airbnb nearly merged with Tujia in 2017 but walked away from the deal involving an equity stake exchange similar to Uber's with former ride-sharing rival Didi in China.

Tujia is well positioned, with 310,000 properties across 255 travel destinations throughout China, and has $300 million in funding from its former parent company Ctrip, China's largest online travel site, and well-known former Morgan Stanley Asia-Pacific research head Richard Ji, managing partner at All-Stars Investment Ltd. Over the past few years, Tujia has acquired the homestay channels of Ctrip and Qunar travel sites, purchased short-term rental platform Mayi.com, and snapped up B2B booking platform Fishtrip to expand into Southeast Asia.

Airbnb's other challenger is Alibaba-backed Xiaozhu, a Chinese homegrown rental-share provider of lower-tier accommodations outside the biggest cities. Xiaozhu too is loaded, with nearly $600 million in venture cash from Jack Ma's Yunfeng Capital, among other high-powered backers. Given the Alibaba connection and its tech knowhow, Xiaozhu is using facial recognition technology to check in guests in China. Maybe this is something that Airbnb could adopt.

The pressure is on Airbnb, a unicorn-valued company that is gunning for an IPO and looking to emerging markets for growth.

Airbnb has doubled its investment in China and tripled its number of staffers in China. But Airbnb's local rivals are spending more in promotions and marketing.

Airbnb China's chairman is realistic about how the American-bred company can move as fast or spend as much money as local competitors. "It is truly eye-opening how much a local company is willing to spend and spend with very little revenue, but based on the idea that it's kind of winner takes all," Blecharczyk told 996 podcast cohosts Hans Tung and Zara Zhang of GGV Capital, an investor in Airbnb. "We probably will never be willing to spend as much as the competitors, but we will definitely spend more than we do in other countries."[13]

Taking Notes on China Business: Evernote

Taking the go-local approach to the extreme, Silicon Valley–based note-taking app Evernote spun out its China unit in mid-2018 and gave the Chinese management team full autonomy over day-to-day operations. About 10 percent of Evernote global sales come from China. In the United States, Evernote has struggled to hold on to its peak popularity of a few years ago as an app to record, organize, and store notes on a smartphone or tablet. What's its formula for success in China?

Evernote was built up in 2007 by Russian emigre and visionary technologist Phil Libin, who was at the helm for nine years before departing to become a venture investor at General Catalyst Partners and later to form AI startup lab All Turtles in San Francisco. When Libin launched Evernote's China-based service in 2012, he gave it a distinctly Chinese name, Yinxiang Biji. But he purposefully made its product offerings no different from the global standard. "We went for a similar core brand value that is not all that different from the way Apple does it. We wanted the product to be so great that everyone wants it, so good that it transcends country borders," Libin told

"We didn't think of China as a country or a market but a place."

Phil Libin
Cofounder and former CEO,
Evernote

me. "We didn't think of China as a country or a market but a place."

Libin traveled to China at least eight times a year to see what was going on with his company's Chinese operations, but he credits Evernote's success to assembling a brilliant team and "making sure they have enough coffee to change the world."

The first general manager at Evernote China was Amy Gu, a Stanford University MBA graduate, founder of two internet startups in China, and a strategy exec at British Telecom Group and China Mobile. She went to work on Evernote China's business development, marketing, product improvement, and customer support, and within four years built the user base to 20 million in China, its second-largest market. Mission accomplished, she left China in 2015 and soon launched Hemi Ventures in the Bay Area to invest in emerging technology companies.

Another anchor in Evernote's success was Ken Gullicksen, a former general partner at Morgenthaler Ventures, who ran corporate development, global sales, and investor relations. He led Evernote's first institutional financing in 2009 and served on its board of directors before joining full-time in 2011, in time to help execute the company's strategy in China. On his watch, Evernote China launched an enterprise version and snared more than $215 million in investment from strategic, venture, and individual investors, including Sequoia Capital China and Morgenthaler Ventures.

One secret to Evernote's success is investor Edward Tian, founder and chairman of China Broadband Capital Partners and a well-known Chinese entrepreneur who took his internet tech startup AsiaInfo Holdings public on Nasdaq in 2001, the first publicly listed high-tech company from China. Tian helped Evernote navigate political waters in China, and he got Evernote running on servers in China at another of his startups, Cloud Valley. The move to a local data center helped speed up connections and data synchs within China and kept Evernote functions working well.

The Chinese name Yinxiang Biji translates well as "memory notes" and, as luck would have it, contains a Chinese character that resembles the elephant shape in Evernote's logo. But Evernote had to downplay social networking aspects of its features in the China version to avoid China's powerful internet censors and privacy issues and to be perceived primarily as an app for personal data storage.

Other smart moves Evernote made: leveraging WeChat to increase awareness; integrating Alipay for payment; localizing its customer service, and launching an e-commerce element specifically for Chinese customers.

The local Chinese management team running Evernote now, led by Raymond Tang, aren't wasting time on carrying this initial success forward as an independent entity. More features are coming specifically for China. New funding was lured in by mid-2018, with ownership split among the startup management locals, Evernote's parent, and investor Sequoia CBC Cross-Border Digital Industry Fund, with Tian and venture *supremo* Neil Shen of Sequoia Capital China. A public offering in China could be coming next. It certainly helps to have connections in China!

VCs Gary Rieschel (r) and Hans Tung sound off at Silicon Dragon in the Valley.

Hurst Lin, DCM Ventures, keeps current by using two mobile phones in China.

Partner Glen Sun picks up honors for Sequoia Capital China as Silicon Dragon VC firm of the year.

VCs Wayne Shiong, James Mi, and Wei Zhou represent a new wave of China venture investment.

Jim Robinson, RRE Ventures, successfully invests in China tech from New York and shares his tales.

Two judges at Silicon Dragon's pitching contest: Brian Cohen, NY Angels, and Hans Tung, GGV Capital.

CEO Robin Li updates the crowd at Baidu World in Beijing about his company's latest initiatives in AI.

CEO Xiaopeng He is driving into the future with his new electric vehicle and smart-car startup, Xpeng Motors.

Gary Rieschel of Qiming Venture gets set to take a victory lap for his venture firm's wins in China.

Richard Liu (l), Morningside
Venture Capital, huddles with Jenny
Lee, GGV Capital, at a recent Silicon
Dragon event in Shanghai.

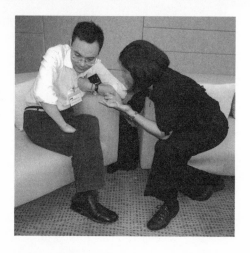

Kai-Fu Lee, Sinovation
Ventures, sees a new world order
for China and Silicon Valley.

VC accelerators Sean
O'Sullivan and William Bao
Bean are putting their stamp
on startups in China and
Southeast Asia.

part two

CHINA'S SILICON DRAGON VENTURE CAPITALISTS

A core group of US-China venture capital leaders set high standards as a Silicon Dragon-style VC market rises to challenge Silicon Valley. A look at who's scoring in China venture investing, and why.

CHAPTER 5

SAND HILL ROAD GETS
A MEGARIVAL

China's red-hot venture capital market has risen to nearly match the US level, and no longer looks to California's Sand Hill Road for cues. Top VC firms in China are funding game-changing innovations at unicorn valuations and achieving high marks for performance.

Sequoia Capital China partner Glen Sun knows what it takes to get ahead in venture capital investing. It's simple. "Make money," he says. "That's it." His remarks on a Silicon Dragon panel I was moderating in Hong Kong drew knowing laughter.

But he and everyone in the audience understands that it's not just that. Sequoia, known for early bets on Google and Apple, has set a high bar with its investments in emerging Chinese companies. With a large team of 60 investors canvassing China for deals, Sequoia is in on just about every hot, up-and-coming Chinese tech company—14 of them went public in 2018 alone, including three of the hottest startups. Sun, who has an MBA and a JD from Harvard, steered Sequoia China's $400 million of investment in Chinese superapp Meituan that netted $4.9 billion, more than a 12-times return based on its eye-popping IPO in September 2018—one of Sequoia's biggest hits.

"We got in early when the space was not so competitive and we've partnered with the best entrepreneurs in China," said Sequoia partner Sun.[1] "As with any good investment company, you have to be performance driven and a meritocracy, and I think making money is the only thing that really matters." Certainly Sequoia, named after the giant trees in the Sierra Nevada range, is a money machine and embraces the slogan made famous by Chairman Mao's successor Deng Xiaoping: "To get rich is glorious."

> *"As with any good investment company, you have to be performance driven and a meritocracy, and I think making money is the only thing that really matters."*
>
> **Glen Sun**
> Partner, Sequoia Capital China

The alpha at Sequoia Capital China is founding partner Neil Shen, a graduate of the Yale School of Management, former investment banker, and a cofounder of Expedia-like Ctrip and budget hotel chain Home Inn. With a net worth of $1.6 billion,[2] Shen is globally recognized as the world's top-ranked venture capitalist on the *Forbes* 2019 list of top 100 VCs[3]—the first time a China-based investor has placed first. A record 21 China venture capitalists made the 2019 ranking, with Sequoia Capital China claiming the most spots, along with Shen and his partners, Steven Ji and Zhou Kui.

Previous top scorers are the Valley's John Doerr of Kleiner Perkins, who aced investments in Uber, Amazon, and Sun Microsystems; Michael Moritz of Sequoia Capital, who scored with Google, PayPal, and Zappos; and Jim Breyer of Breyer Capital, who made a fortune on Facebook.

Under Shen's leadership, Sequoia Capital China has been on a winning streak with several of the tech startup champions featured in this book. Sequoia avoided large flops that got funded in a rush of irrational exuberance, like bike-sharing startup Ofo. Both Shen and Sun have been well schooled in the craft of smart venturing, applying mathematics to engineering, business, and computer science fields and using their gut instincts.

Sequoia Capital is best known for its successes with Apple, Google, and Oracle, all within the Silicon Valley orbit and dating back a couple of decades. US partners Michael Moritz and Doug Leone had the foresight to extend the franchise to China in 2005, when Sand Hill Road was just discovering Chinese tech ingenuity. Sequoia raised an initial $250 million fund for China. The partner team of Shen along with Zhang Fan, who was recruited from DFJ ePlanet Ventures, quickly invested in 27 would-be digital stars in China. A setback came when Zhang, known for his early bet on search engine Baidu, resigned in 2009 for "personal reasons" amid allegations of bribery and kickbacks.[4]

While that bruised Sequoia's stellar reputation, US partner Moritz, a big fan of China's work ethic and entrepreneurial culture, helped the Chinese affiliate reset. Today, with nearly $20 billion to invest in China, Sequoia Capital China is the largest Valley-anchored firm prospecting in the Middle Kingdom and has an unbeatable track record and commitment. Nearly two-thirds of Sequoia's new mega $8 billion global fund is reserved for China. Additionally, Sequoia's three recently launched funds totaling $2.5 billion[5] are ready for China's emerging high-tech companies. One secret to Sequoia's success is the Yale Center Beijing, which was established by Sequoia China cofounder Shen and his alma mater to run a one-year leadership education program for Sequoia-funded founders.

This is a pivotal moment. Never before in the ebb and flow of venture's 50-year history has China risen to the US premium level. Now, in many measures—capital under management, investment totals, fund performance, unicorns, breakthrough portfolio companies—it does.

"China's emerging innovations and the scale of its venture capital markets are driving China's competitiveness on the world stage," says Gary Rieschel, founding managing partner of leading US-Chinese venture capital firm Qiming Venture Partners. Powerful indicators point to the center of gravity shifting from longtime leader Silicon Valley toward China's startup ecosystem.

Consider:

- China venture spending almost reached the US level in 2018, soaring 56 percent to $105 billion compared to a 42 percent gain to $111 billion for the US.[6] In the first half of 2018, China expenditures surpassed the US.[7]
- China VC spending was a mere $5.6 billion in 2010, over-shadowed by the United States at $35.3 billion.
- Of $275 billion of venture investments globally in 2018, China and the United States were just about even, but in 2010 China's share was only 12 percent, eclipsed by the United States at 66 percent of $47 billion invested worldwide.
- China venture fund-raising climbed about 8 percent to $23.6 billion in 2018 and gained for most of the past eight years to a peak in 2016 of $44.7 billion—skyrocketing from just $9.9 billion in 2010. Comparatively, US venture funds reached $48.8 billion in 2018, up from $16.6 billion in 2010.
- An analysis of China and US venture fund performance reveals that prominent China VC funds earned an average 21.4 percent return, slightly higher than most US funds.[8,9]
- Of the world's 40 best venture capital bets of all time, 12 are Chinese tech startups.[10]
- China inked seven of the world's top 10 venture deals in 2018, including the largest ever—the gigantic $14 billion funding of Jack Ma–controlled Ant Financial.[11]
- China weighed in with 86 unicorns in 2018, topped only by the United States with 151. China claimed one-quarter of 311 venture-backed privately held companies in the global unicorn club of startups valued at more than $1 billion.[12]
- Chinese AI-powered digital content app ByteDance ranked number one as the world's most valuable unicorn at $75 billion and bypassed Uber at $68 billion. On-demand Chinese ride service Didi placed third at $56 billion.

- Out of 190 US IPOs that raised $47 billion in 2018, China scored 31 IPOs that pulled in $8.5 billion, up from 16 at $3.3 billion in 2017—an eight-year high[13] since Alibaba's mega $25 billion IPO in 2014. Four of the top US IPOs in 2018 were from China.[14] Additionally, 44 Chinese companies went public in Hong Kong in 2018 and attracted $32 billion, nearly triple the level a year earlier.[15] See tables 5-1, 5-2, and 5-3.

Table 5-1

China's Venture Investments Soar

	2018	2015	2010
China	$105 billion	$44.7 billion	$5.6 billion
United States	$111 billion	$75 billion	$30.8 billion
Global	$274.6 billion	$155 billion	$46.4 billion

Source: Preqin

Table 5-2

China's Venture Fund-Raising Climbs

	2018	2015	2010
China	$23.6 billion	$22.7 billion	$9.9 billion
United States	$48.8 billion	$34.6 billion	$16.6 billion
Global	$79.1 billion	$70.1 billion	$31.4 billion

Source: Preqin

Table 5-3

4 Chinese IPOs in Top 10 US Listings in 2018

	Amount Raised	Rank	Exchange
iQiyi	$2.4 billion	3	Nasdaq
Pinduoduo	$1.7 billion*	4	Nasdaq
NIO	$1.2 billion	9	NYSE
Tencent Music	$1.1 billion	10	NYSE

* total value includes overallotment

Source: Dealogic

China is undergoing an economic revolution driven by technology innovation and venture capital on a scale the world has never seen before. This is China's first generation of Chinese tech entrepreneurs and VCs since the Cultural Revolution and since reforms opened up the country to private enterprise. In less than two decades, China has measured up to Silicon Valley. It was 50 years ago that Sand Hill Road–style investing was pioneered by Don Valentine of Sequoia Capital and Tom Perkins of Kleiner Perkins, when the Valley was still dotted by fruit trees and when China was largely agricultural.

I followed the venture trail from Silicon Valley to Beijing and Shanghai and tracked China's ascent. Today, investment returns of top-tier Chinese venture funds outdo most US funds and also real estate and stock market investments—although venture investing is riskier and highly cyclical, and the money is not liquid.

Three Digs to Find Gems

Venture capitalists in China have upgraded over the past 20 years, in three stages.

The first was to invest in China copycats of Amazon, Facebook, Google, and YouTube, and manage deals by parachuting in from Silicon Valley. The second was on-the-ground China teams funding local entrepreneurs with customized ideas made and designed for China tastes. Third is to move to full partnerships investing in disruptive Chinese startups that have the potential to be copied outside China and go global.

In synch, China tech has progressed through three types of technology phases: internet startups from 2003 to 2010, mobile-centric startups for the next few years, and, today, advanced technologies and business models in artificial intelligence, biotech, self-driving, robotics, drones, livestreaming, mobile payments, social networking, and social commerce that impact broad sweeps of the economy such as transportation, finance, health care, and education. These deals are not the low-hanging-fruit deals that venture capitalists have been known to favor. It's a stretch that would have been unimaginable even just a decade ago.

"Chinese tech companies won battles in almost all of the sectors, and China's internet market became a stand-alone and self-sufficient market, with everything from search, to content, to e-commerce," says David Yuan, founder and managing partner of Redpoint China Ventures, an offshoot of Silicon Valley's Redpoint Ventures. "In mobile and consumer internet, Chinese tech innovations have become advanced and world leading due to the size of its domestic market. Starting 2015 to now, increasing numbers of Chinese tech companies are venturing outside of China to pursue their global ambition, and the trend will continue for many years to come."

From New York City, venture investor Jim Robinson weighs in. "It was true 25 years ago that China was copying, but today China has cutting-edge tech. At Tsinghua University, it's as good as it gets," says Robinson, general partner of New York–based RRE Ventures, a longtime investor in China who sees the Chinese coming up fast in quantum computing, machine learning, blockchain, and software for gaming.

But there's a potential downside. Many venture-funded Chinese startups are burning cash like there's no tomorrow to chase growth, well before profits (just as Amazon did).

> *"Chinese tech companies won battles in almost all of the sectors, and China's internet market became a stand-alone and self-sufficient market."*
>
> **David Yuan**
> Founder and managing partner,
> Redpoint China Ventures

> *"It was true 25 years ago that China was copying, but today China has cutting-edge tech. At Tsinghua University, it's as good as it gets."*
>
> **Jim Robinson**
> Cofounder and general partner,
> RRE Ventures

"We believe the era of large-scale profitability of Chinese start-ups has just begun," says Wei Zhou, founder and managing partner of Beijing-based China Creation Ventures, which he spun out with a team from Kleiner Perkins China. "I never worry about a startup losing money. I worry more about how much value it is creating. When they win, they will make a lot of profit."

Megafunds are chasing tomorrow's China stars in a tech economy that keeps churning in spite of the larger issues of US-China trade wars and technology leadership races. Competition is heated to invest in young, techno-based businesses, and unicorn-valued Chinese startups keep multiplying. The rallying cry is, "Don't miss the next new thing in China." Social commerce innovator Pinduoduo came out of nowhere and caught many venture investors by surprise, and they missed it.

> "We believe the era of large-scale profitability of Chinese startups has just begun."
>
> **Wei Zhou**
> Founder and managing partner,
> China Creation Ventures

China's venture market has narrowed to a core group of longtime successful firms. Pioneering China investors Kleiner Perkins and Draper Fisher Jurvetson no longer invest there. New independent shops have spun out from Silicon Valley firms: China Creation Ventures from Kleiner Perkins, Long Hill Capital from NEA, and 1955 Capital from Khosla Ventures.

More capital is flowing to proven Chinese venture firms from large US pensions such as CalPERS (California Public Employees' Retirement System) and CalSTRS (California State Teachers Retirement System); sovereign wealth funds, including Singapore's Temasek and GIC; rich serial entrepreneurs; family offices; fund of funds; and university endowments, including Yale, Princeton, Northwestern, and Duke. Money also has flowed from Chinese government–backed investor groups into Chinese currency RMB funds that can invest directly in Chinese startups and take them public in China, such as the Nasdaq-style ChiNext in Shenzhen and[16,17] the new Shanghai Science & Technology Innovation Board for China's emerging companies to list. But tighter risk controls to deleverage the domestic financial industry in China and poor performance of first-time Chinese fund managers have made it tougher to raise yuan funds. Most US-anchored firms today investing in China have funds in both renminbi and US dollar funds, which typically fund Chinese startups through offshore-owned subsidiaries.

Angel funds have sprung up from wealthy successful serial entrepreneurs like Alibaba's Jack Ma, Xiaomi's Lei Jun, and Zhen-Fund from angel investor Bob Xu, the founder of NYSE-listed New Oriental Education & Technology Group.

There's an ever-growing pot of gold for the best-of-breed investors to back China's emerging tech businesses.

Table 5-4

Major Sino-US Venture Power Players

FIRM	Capital	Funds	Investments	Exits	Launch
Sequoia Capital China	$20 billion	21	500*	70	2005
Qiming Venture Partners	$4.3 billion	12	344	50	2006
GGV Capital**	$6.2 billion	13	345	103	2000
DCM Ventures***	$4 billion	14	400	75	1996
Matrix Partners China	$3.5 billion	10	520	65	2008
GSR Ventures	$2.1 billion	12	255	26	2005
Sinovation Ventures	$2 billion	7	350	40	2009
Lightspeed China	$1.5 billion	7	96	15	2012
Redpoint China	$1 billion	6	70*	9	2016
ZhenFund	$1 billion	5	700	45	2011
NEA	—	8	50	20	2003

Sources: Silicon Dragon research, VC firms

* companies
** includes China and United States
*** includes China, United States, and Japan
—NEA invests out of one fund, with no separate China fund.

Table 5-5

Key China Unicorns and Chief US-China Venture Investors

Xiaomi:	Morningside Venture Capital, Qiming Venture, IDG, Temasek, DST Global
Meituan Dianping:	Qiming Venture, Sequoia Capital China, General Atlantic
Pinduoduo:	Lightspeed China, IDG Capital, Banyan Partners, Sequoia Capital China
ByteDance:	SIG Asia, GGV Capital, Qiming Venture, New Enterprise Associates, Hillhouse Capital, Sequoia Capital
Kuaishou:	DCM Ventures, Sequoia Capital China, Morningside Venture Capital
SenseTime:	Tiger Global, CDH, IDG Capital, Fidelity International, Temasek, Silver Lake, HOPU Capital
Face ++:	Qiming Venture, Sinovation Ventures
Didi:	GGV Capital, GSR Ventures, Matrix Partners China
DJI:	Sequoia Capital China, Accel Partners
NIO:	Sequoia Capital China, Temasek, IDG Capital
Xiaohongshu:	ZhenFund, GGV Capital
UB Tech:	Qiming Venture, CDH Investments
Mobike:	Qiming Venture, Vertex Ventures, Hillhouse Capital

Source: Silicon Dragon research

Stumbles along a Digital Silk Road

Despite the growth opportunity for startups, the wide-open land-scape for new technologies, and the entrepreneurial culture of China, success is not a given. Several top-tier Sand Hill Road firms have stumbled along this digital Silk Road, not able to transfer their knowledge in from the Valley.

Staying power in China is tough. Tim Draper, known for his bets on Skype, Hotmail, and Tesla, was a very early supporter of China tech

innovation, dating back nearly two decades. But he's stopped investing in China, though an early bet on search company Baidu through
DFJ and ePlanet made a 33 percent return on a 28 percent stake after
the search startup went public on Nasdaq in 2005. Draper faults the
difficulty of getting money out of the country due to China's tightened
controls over capital going outside its borders. Now Draper, founder
and managing partner of Draper Associates and the superhero behind
entrepreneur school Draper University in San Mateo, keeps a hand in
China as a limited partner investor in a Draper Dragon fund raised in
2016 and run by Andy Tang, the CEO of Draper University.

Likewise, venture investor Robinson at RRE Ventures relates a
story of trying to get money out of a China deal where millions of
dollars were made from a sale to Chinese conglomerate Ping An. He
first tackled a tax issue where six provinces were each requiring 20
percent of the total return. Once that was solved with the help of a
Chinese law firm, Robinson had to figure out how to get the renminbi out. Some laughable suggestions were that he should buy a
large diamond and bring it home or take a duffel bag of cash to Hong
Kong and trade it for HK dollars, but his firm hit upon a solution to
route through two wholly owned foreign subsidiaries, in Hong Kong
and in the Cayman Islands.

Charles River Ventures was led to China early on by then-partner
Bill Tai, who had established the Boston firm's Silicon Valley office.
In 2006, Charles River backed an innovative, customizable web
browser, Maxthon, based in Beijing, but Maxthon never got into the
major leagues. Tai transitioned to a partner emeritus at CRV in 2014
and stepped off the board of Maxthon in 2016. Now a prolific individual investor, enthusiastic blockchain investor, and kite-boarding
athlete, Tai recently saw an early bet on China-linked video confer-
encing software company Zoom pay off with an IPO on Nasdaq. The
Mayfield Fund got into China in 2004 through a partnership spearheaded by former managing director Kevin Fong with Beijing-based
GSR Ventures, which Fong advises. Mayfield is investing in the

United States and India instead of China today. Matrix Partners got a China edge in 2008 when former eBay CEO Bo Shao, founder of eBay-acquired EachNet, and a team of venture capitalists from long-time China investment firm WI Harper split to form Matrix China Partners. Accel Partners ventured to China in 2005, when Jim Breyer teamed up his former firm with Beijing-based IDG Capital, where he is now cochairman. IDG Capital brought the first foreign capital into China in 1992 and counts huge returns from early investments in Tencent, Baidu, and Xiaomi.

A few cutting-edge Bay Area firms never mustered the nerve or the urge to invest in China but have kept a close watch. Andreessen Horowitz partner and widely followed digital China expert Connie Chan tracks China's progress and seeks opportunities for the firm's US startups. Peter Thiel's Founders Fund appointed Jeff Lonsdale in 2018 as managing director–Asia, and he's traveling often to Asia from Silicon Valley to see what's going on and scout research opportunities for investment. Khosla Ventures, founded by Indian-born billionaire Vinod Khosla, seeks to help US portfolio companies get traction in China but hasn't invested directly. Several Silicon Valley firms still dismiss venture investing in China as too risky, and many in the clubby Sand Hill Road community prefer a comfort zone within a close radius of the Bay Area, fearing the unknown and language and cultural barriers.

Experienced China-side venture investors have sometimes struck out. Matrix Partners China, GSR Ventures, and ZhenFund—as well as Alibaba—got caught in China's bike-sharing craze and meltdown after injecting $2.2 billion into Chinese bike-sharing startup Ofo, which has crashed after so much hype, a victim of overexpansion, competition, and a money-losing business model. Another China deal that cost its investors dearly was failed blogging platform Bokee, funded with $10 million from SoftBank, Bessemer Venture Partners, and Granite Global Ventures (now GGV) a decade ago. Bokee's founder, Fang Xingdong, had a good grasp of digital media content but had conflicts with his board and lost the CEO seat after hiring hundreds of employees who had to be laid off when online advertising revenues didn't keep pace. Bessemer closed its China shop after that.

Trekking on the Venture Trail

China's history as a budding venture superpower can be traced back to Silicon Valley when Sand Hill Road firms began looking for the next big thing after the US dotcom bust in 2002. Founders of Asia-oriented venture firms Ta-lin Hsu of H&Q Asia Pacific and Lip-Bu Tan of Walden International had already paved the way and made it easier for Valley leaders to get comfortable with the idea of investing in China. Valley adventure capitalists got an initiation to China in 2004 when Silicon Valley Bank led a one-week whirlwind tour of Beijing and Shanghai for honchos John Doerr of Kleiner Perkins, Don Valentine of Sequoia Capital, and Dick Kramlich of NEA, among others, to meet with entrepreneurs, public company executives, and government officials. They came back in awe of the market potential and the Chinese entrepreneurial machine. Soon, partners at these Northern Californian venture firms were using guest offices at Silicon Valley Bank in Shanghai as their base. Within a few years, many Valley firms established offices in Chinese cities, debuted China-specific funds with local China partners, and inked deals.

Considering that American venture leaders from Silicon Valley who entered China were entirely new to the People's Republic, didn't know the landscape, weren't familiar with the culture, didn't speak the language, and were investing in untested Chinese entrepreneurs, it's pretty remarkable that several firms have achieved good results.

In the early days of venturing in China, led by Silicon Valley firms, several missteps were made. Small teams were stretched to cover the opportunities, partners parachuted in from California couldn't fill the void, teams hastily put together in new venture outposts clashed, and strategic decisions and tactical moves were called from Silicon Valley rather than the front lines.

I remember when Kleiner Perkins founder John Doerr blasted into Beijing and held a press conference in 2007 to launch the US firm and duplicate its successful Valley formula in China. A $360 million fund was raised specifically for China. Two well-regarded Chinese venture capitalists, Tina Ju and Joe Zhou, were handpicked to head up Kleiner Perkins in China. But Zhou bailed one year into the partnership, in early 2008, over incompatible deal-making styles. Zhou bought out his seven portfolio deals from Kleiner China and pulled them into a new firm, Keytone Ventures, in mid-2009, with many of the same limited partner investors who had backed Kleiner's inaugural effort in China. Meanwhile, Kleiner Perkins scrambled to stabilize and build up the team and work through investments that didn't work out because of poor timing or execution or inexperienced entrepreneurs who had never run a business before. Kleiner Perkins wound down its China activities after a team of investors departed in 2017 to form the new early-stage technology investment firm China Creation Ventures, operated in China for China. Kleiner Perkins is left with a skeleton crew in Shanghai to work on past investments. The departure of venture capitalist Mary Meeker from Kleiner Perkins in 2018 was another blow for any further plans for China. Meeker's annual tech trends report under the Kleiner banner was bullish on China. Meeker is now investing from a fund for growth investment at her new firm, Bond.

Connecting in Chengdu

Venture investing requires identifying, researching, and funding passionate founders who dream big, innovate huge, and compete to win with the right team, the right sector, and the right business model. It's a tall order. Venture capital is hard enough in Silicon Valley, but the usual work of researching promising sectors, finding breakaway startups, getting referrals, checking references, and diving into bookkeeping records is more painstaking in China.

Unearthing gems requires going outside well-traveled paths. I was in Chengdu a few years ago to interview a budding tech entrepreneur and bumped into venture capitalist Richard Liu of Hong Kong–based Morningside Venture Capital who was just wrapping up a meeting with the same founder. This serious-minded and time-consuming approach is what it takes, and it's why Liu in his 15-year career at property tycoon Ronnie Chan's Morningside has become one of China's most successful venture investors. Liu made early, gutsy bets on smartphone maker Xiaomi and social entertainment company YY that paid off handsomely when both Chinese startups went public.

Wins come from exits: when venture investors cash out their holdings in startup companies in public offerings or from acquisitions. The conventional wisdom is most new startups fail, as many as 90 percent, and that only a small percentage of venture firms make money from their investments. Another common rule is that 80 percent of investment returns come from 20 percent of startup deals—known as the "Babe Ruth effect" for the American legendary baseball player who struck out a lot but also scored many home runs. In venture capital, one successful deal can make up for the losses and "return the fund," or bring in the investment returns needed for the firm to profit. It usually takes 10 years for a fund to be fully invested. Good results are hardly guaranteed. Like fine wines that vary according to vintage year and can improve with age, venture capital returns fluctuate from boom to bust cycles and by year of the fund's original investment.

Giant NEA Steps In

New Enterprise Associates (NEA), one of Silicon Valley's largest and longest-established firms, was a pioneer in China and managed through some rough spots. The firm now has a portfolio of more than 20 companies but initially found the going more difficult than originally anticipated, as managing general partner Scott Sandell once told me when he was leading NEA's China investing. NEA began investing in China very early, in 2003, when most Silicon Valley players were not about to leap in. The firm made money on bold investments of $120 million in Chinese semiconductor maker SMIC, which was championed by leading Asian venture capital firms H&Q Asia Pacific and Walden International, and helped NEA to get in on this deal. The chip maker, which I toured in Shanghai with then-CEO Richard Chang, who was recruited from a leading Taiwan foundry to run operations, went public in 2004 and raised $1.7 billion on the NYSE. SMIC has become one of the world's largest chip makers but lags more advanced foundries in Taiwan, the United States, and South Korea. Others in NEA's China portfolio required more work to get to an exit, and the firm briefly paused funding more startups in 2007 to work through a backlog. When NEA wanted to place a US-based partner in China, and partner Sandell opted to continue shuttling to China rather than move with his young school-age children to Shanghai, the firm's legendary founder Dick Kramlich stepped in. Wanting to show the firm's commitment to China, he and his wife, Pamela, left their Nob Hill home behind and moved to Shanghai in 2008 to oversee the firm's China investments for a year and a half. Kramlich worked with local partner Xiaodong Jiang, the firm's first full-time employee in China, who opened both its Beijing and Shanghai offices and led the on-the-ground investing team for 11 years. In 2016, Jiang left to form a spin-out firm, Long Hill Capital, raising a $125 million fund mainly for health-care companies with NEA as an investor and adviser. NEA has continued to pursue China deals, co-investing with local independent teams and funding startups from Silicon Valley funds. NEA has powered up its China activities from the Bay Area by adding former corporate and securities lawyer Carmen Chang as a special adviser in 2012, who soon moved up to partner. Now a general partner and chairman and head of Asia, Chang brings deep knowledge to the firm as a former partner at law firm Wilson Sonsini Goodrich and Rosati (WSGR), where she was involved in many seminal trans-actions, such as the IPOs of NEA-backed Chinese semiconductor companies SMIC and Spreadtrum. In 2018, NEA came out swinging with a co-investment in ByteDance, one of China's hottest startups, and an outlay for a promising Chinese startup in online education, Zuoyebang, with Sequoia Capital China and GGV Capital.

China's Buzzing Hubs

Throughout, the chess board of China tech and venture has con-tinued to evolve. The action has spread from leading hubs Beijing and Shanghai to Shenzhen, home to Tencent and a capital of hard-ware innovation, and Hangzhou, headquarters for Alibaba. Farther out, Chengdu, Wuhan, and Chongqing have gotten on their radar. Hong Kong, known for trading, finance, and real estate, has become a magnet too for venture capitalists and startup founders who like the city's tax incentives, pleasant lifestyle, tech-oriented universities, proximity to Mainland China, and new tech-friendly reforms to list on the Hong Kong Stock Exchange, plus an institutional investor base that's familiar with Chinese business models.

China's hubs are fed by engineering and managerial talent; clus-ters of coworking spaces and incubators such as hardware accelera-tor HAX; loads of networking events; and major professional firms that provide young companies with banking, legal, and accounting services. China counts 7,500 incubators and maker spaces, the most of any nation.[18]

While China's venture and tech startup scene buzzes, Silicon Valley has not lost out as the world's center of technology innovation yet. The Valley remains, for now, the crossroads between East and West that draws venture seekers from all over the world.

The United Airlines lounge at San Francisco airport is a popu-lar meeting point for these frequent US-Asian business travelers to exchange information and relax before a long flight—and I can't go incognito there, since many of my venture capital connections are hanging out there too and may be on the same flight with me. It's not uncommon for China-based venture capitalists to own a second home in wealthy Atherton, Los Altos, or Palo Alto to keep a pulse on what's going on in the Valley and spend family holidays. It's also a hedge in case something goes wrong in China or the harsher life-style in China's crowded cities becomes too much.

But don't count on China venture capitalists to bolt from their home country. It can be very exciting and rewarding to be on the front lines of this huge transformation in China. The pace is incredibly fast. Ideas travel quickly, often fed by WeChat groups. Entrepreneurs in China regularly put in 12-hour workdays, 6 days a week, or 996, as the phrase goes that has caught on in China tech and venture circles. It's nothing for entrepreneurs and venture capitalists in China to have midnight conference calls, sometimes because they're dealing with partners or colleagues in a 15-hour-difference time zone away in California.

Venture partners from the top firms investing in China are an elite bunch. They sport business and engineering degrees from top-notch places like Harvard, Stanford, Yale, Princeton, Northwestern, UC Berkeley, Wharton, or Cornell and company operating experience from multinationals Google, Qualcomm, Intel, and Microsoft. They work well with teammates and bring passion about leading-edge innovations to the job. For many, it's not just about the money. Some are also techie geeks. Most are internationally minded, spend a lot of time on the road, like to travel, and don't mind the red-eye flights too much, at least in business class. They work long hours but enjoy fine wines and beach resorts and have more than enough money to pay tuition for their kids to go to the best private schools. They are chauffeured around by private drivers or Uber (Didi in China) and work in beautifully designed and spacious offices along Sand Hill Road close to Stanford University or city center towers in Beijing and Shanghai. It may sound cushy, but it takes hard work, commitment, and determination.

Crosscurrents and Synergy

China's Silicon Valley owes much of its initial magic to a reliance on the United States for a cross-border flow of ideas and capital. A two-way highway runs from Beijing's Zhongguancun Software Park to Menlo Park's Sand Hill Road, raising capital and funding startups hinged from both coasts of the Pacific Ocean. This two-way channel creates synergy and speeds up startup launches, innovation, and scale across the United States and China, as well as globally.

The tech investing pipeline from China into the United States has been increasing, even though tensions from Washington are rising. China venture funds invested $3.1 billion in the United States in 2018[19]—mostly in emerging deep-tech, nerdier but critical fields, up from $2.1 billion in 2017—and almost zero in 2010. Additionally, China co-invested with US venture firms in 231 deals in 2018, holding steady at about 9 percent of total US deals.[20]

This level of venture flows caught the attention of the US Department of Defense and its Defense Innovation Unit, charged with helping the military make better use of emerging technologies. The unit published a report about China technology transfer practices citing statistics that underscored how embedded the Chinese are in US technology investments.[21]

> "We are investing in cross-border even though there really is not a border for these innovations."
>
> **Ming Yeh**
> Founding managing partner, CSC Upshot Ventures

Although funding by China venture firms into the United States held up and even grew in 2018, the impact of deteriorating US-China relations over trade and tech leadership issues and the related late-2018 arrest of the CFO of China's telecom giant Huawei over possible Iran sanction violations, did cause corporate venture investment from Chinese companies into the United States to slow down.[22] Another dampening factor on the corporate VC flows was more stringent controls by the Chinese government on capital outflows.

Longtime cross-border investors from Silicon Valley downplay impact from US-China tech and trade tensions and regard it as a political issue that doesn't impact their tech economy world much. "Technology is global by definition, and it's without borders," said Ming Yeh, founding managing partner at CSC Upshot Ventures in Silicon Valley. "We are investing in cross-border even though there really is not a border for these innovations," she said at a recent Silicon Dragon forum in the San Francisco Bay Area. David Lam, general partner at cross-border tech investment firm Atlantic Bridge, pointed out that any barrier is largely caused by politics. "Global supply chains are international, and the movement is unstoppable. It's already happened. So it's hard to turn back the clock. Some of the political leaders don't necessarily share that view. They are the ones driving a lot of that policy and that does obviously create barriers, but they are more politically motivated than business motivated."

> "Global supply chains are international, and the movement is unstoppable. It's already happened. So it's hard to turn back the clock."
>
> **David Lam**
> General partner, Atlantic Bridge

A US-China Venture Bridge Detours

But the rifts are causing the bridge between Silicon Valley and China cross-border investing to be rerouted. Instead of a two-way road on one level, venture traffic is starting to split into two avenues going in a one-way direction. The result? Separate Sino-US venture spheres forming within their own geographies. In one sign of the change, Beijing-based Sinovation Ventures is refocusing its prior two-pronged US and China deal making to concentrate on

China-side deals, with chairman and CEO Kai-Fu Lee leading the firm's stronghold of artificial intelligence in Beijing.

Against this backdrop, plenty of new funds directed to China are still being launched. See table 5-6. Chinese investment firm Hillhouse Capital, which has backed many iconic companies including Tencent, Baidu, and JD.com, raised a mega $10.6 billion fund in September 2018 to invest in China and in Asia.

Table 5-6

New China Funds from Sino-US Venture Firms: 2018–2019

Sequoia Capital China	$2.5 billion	3 Chinese funds, raising a new $8 billion global fund
Qiming Venture Partners	$1.39 billion	3 funds including RMB fund of $340 million*
GGV Capital	$1.88 billion	4 funds plus RMB fund of $250 million*
Sinovation Ventures	$500 million	1 fund plus RMB fund of $375 million*
Matrix Partners China	$750 million	1 fund
Lightspeed China Partners	$560 million	2 funds
China Creation Ventures	$200 million	1 fund
Long Hill Capital	$265 million	1 fund
DCM Ventures	$750 million	1 fund
ZhenFund	$190 million	1 fund
Redpoint China	$400 million	2 funds

*RMB fund sizes are cited in approximate US$ value

Sources: Silicon Dragon, venture firms

Long Hill Capital, the spin-off from NEA, launched a second fund of $265 million to invest in health care and consumer businesses that serve the aging population in China. China Creation Ventures, the spin-off from Kleiner Perkins, collected $200 million for a new fund of $200 million in 2018, bringing its capital to more than $400 million to invest in Chinese tech, telecom, and media startups. His firm scored its first IPO in late 2018, the Hong Kong listing of Chinese startup Wanka Online, an alliance of Android smartphone brands in China focused on increased traffic and commercialization. "We want to better capture the opportunity in China, because China innovation moves so quickly," says Zhou, who has a team of 16 partners. "There is always something surprising about China."

Another offshoot, 1955 Capital from prior Khosla Ventures investor Andrew Chung, launched a $200 million fund to focus on helping Western entrepreneurs commercialize their technology in China and India.

Center of Gravity Shifting Eastward

A next frontier for venture investors has opened in Southeast Asia, which is feeding startups with a large and fast-growing population of 665 million and a big internet base of 260 million. Business models and innovations transfer easily from China to this region and catch on quickly thanks a cultural similarity and an influx of capital from China and regionally. Funding for Asian companies climbed 11 percent to $81 billion in 2018 from the previous year—a major leap from $7 billion in 2013. The number of deals jumped 42 percent to 5,066 in 2018, the fifth consecutive year of increases.[23]

Though I'm familiar with Southeast Asia from travels and interviews in the region, I was still amazed by the budding entrepreneurial culture at a huge and flashy tech summit in Bangkok, where I was a speaker. The event, run by local tech and business communications group TechSauce, drew hundreds of startups and venture investors from nearby hubs and some from the West. It can make the annual TechCrunch Disrupt conference in San Francisco seem small and not nearly so dazzling by comparison.

Southeast Asia is "another China-sized opportunity," contends Nazar Yasin, founder of investment firm Rise Capital, which funds expanding internet businesses in emerging markets. Speaking at a recent Silicon Dragon forum in Hong Kong, investor Yasin pointed out that Southeast Asia startups accounted for 5 of the 12 largest internet financings globally in 2017 and three of the largest ones globally in 2018. Several China-oriented firms have opened offices in Singapore, including GGV Capital. Gobi Partners, formed in 2002 and named after the desert that stretches across northern China, runs the Alibaba-linked Hong Kong Entrepreneurs Fund but has turned to the Southeast Asian market too. Gobi has invested in 58 startups in the region and launched a $10 million fund for early investments in Indonesia and a $14.5 million fund for Malaysia. I profiled several young superstars for *Forbes Asia's* "30 under 30" special report, many of them from Indonesia, Malaysia, and Singapore, and they are just as impressive as the Chinese entrepreneurs who already have shown the way.

The Venture Gold Diggers in China: Who's Who

Despite the appeal of new regions, China's size, entrepreneurial culture, tech talent, and pace continue to make the Middle Kingdom the beacon in Asia. The venture firms that put down deep roots by hiring Mandarin-speaking partners and opening offices in China showed they were serious. They have persevered through various cyclical downturns and upturns common to venture investing, and they continue to bulk up.

The following section looks more closely at the Sino-US venture investors who are most active in China and who caught China's tech boom at the right time when the market was young, the opportunities seemed boundless, and the valuations to invest were a lot less than they are today.

The Dawn of Qiming

One of China's longtime premier venture firms, Qiming Venture Partners, has primarily focused on China-side deals from five offices in China and has more recently opened a US office and launched a US fund. Qiming got an early start in China tech investing in 2006, spearheaded by founding partner Gary Rieschel, a pioneer and mentor in the business of venture—and someone whose career I've followed since before the dotcom bubble burst. Rieschel had a good sense of an about-to-boom tech economy from working in Japan during the late 1980s and from setting up and running SoftBank's US venture capital business from Silicon Valley in the go-go period of mid-1990s through the downward cycle ending about eight years later. By 2005, when Rieschel had wound down his involvement with SoftBank, he and his wife moved to Shanghai just as internet startups were being seeded. He formed Qiming Venture in 2006 with Duane Kuang, the former director of Intel Capital China, in partnership with a venture capital firm in the Seattle area, Ignition Partners. But Rieschel decided later that Qiming should have its own identity rather than using US brands to build the business in China. Rieschel wanted to select a Chinese name for the firm, and his wife came up with Qiming, which means "to inspire" or "to enlighten," from the names of their two Chinese-named children.

Rieschel once said he would run a victory lap from success in China, which he's seen evolve to the point that it's increasingly challenging the US for tech leadership. He's running that lap now as some Qiming funds bring in investment returns of around 30 percent in the top echelon of venture firms investing in China.

Qiming manages more than $4.3 billion across 12 funds—seven US dollar funds and five RMB funds—and has invested in more than 280 young, fast-growing, and innovative companies in clean tech, health care, the internet, and information technology. More than 50 of them have been acquired or gone public in the United States, China, or Hong Kong, including smartphone maker Xiaomi and on-demand food delivery and services app Meituan.

Unlike many other venture firms, Qiming partners specialize in sectors. For instance, Hong Kong–based managing partner Nisa Leung, who previously worked with Rieschel at SoftBank, specializes in health-care investing.

Of the firm's five managing partners, Qiming managing partner JP Gan is on the *Forbes* list of top VCs globally, ranked fifth thanks to his investments in Meituan and popular selfie-editing app Meitu. Some 20 of the firm's portfolio companies have grown up to become unicorns. Others have been acquired in big deals, such as bike-sharing upstart Mobike, absorbed by Meituan in 2018.

After 11 years living in Shanghai and working from the firm's offices in Jin Mao Tower, Rieschel moved back to the US in 2016 and settled in Seattle. In 2018, he launched a US-focused fund of $120 million, the firm's first outside of China, investing in emerging companies in health care from offices in Seattle, Palo Alto, and Boston. Rieschel continues to be actively involved in overseeing what's going on in China.

The Granite in GGV Capital

GGV Capital's new funds totaling nearly $1.9 billion in 2018 and its $6.2 billion capital under management show that its US-China cross-border investing strategy must be working. The firm focuses on e-commerce and mobile internet deals. GGV Capital's reputation was built on an investment in Alibaba in late 2003, when that wasn't a sure bet, but it hedged that risk by following initial investments from Goldman Sachs and SoftBank a few years earlier. GGV ended up putting $7.8 million of investment in Alibaba and claims to have made $200 million on its investments, although the firm missed a bigger prize by selling most of its stake in Alibaba before the mega IPO in 2014.

Since then, GGV—originally called Granite Global Ventures but changed to GGV in 2008—has gone on to seek out what could be the next Alibaba. The firm has invested in several deals that were seeded or founded by angel investor and Xiaomi CEO Lei Jun. GGV partners regularly go out to seek startups from entrepreneurs at Stanford, Harvard, and other fine universities. The firm has come into the spotlight, a long way from the four founding partners, including former Silicon Valley lawyer Joel Kellman and Singapore-based high-tech investor Thomas Ng, who all have left the firm to retire or become personal investors. I remember meeting some of them at a TDF Ventures annual meeting in 2000 at the Four Seasons Hotel, where we heard several young tech entrepreneurs tell about their startups. We toured the Xintiandi high-end shopping and residential zone that was just opening and met with its developer, Vincent Lo from the Shui On Group. It was an exciting, heady time.

GGV Capital has gone on to invest in 51 market leaders in the United States and China—and some others on the way up. One is Chinese drone maker EHang, which wowed the crowds at the Consumer Electronics Show a few years ago with its flying taxi concept but has since retrenched to the China market, where it is rising again. On the upside, about half of those 51 GGV portfolio companies have gone public. The cross-border investment firm boasts a 25 percent internal rate of return (IRR), a measure of investment performance, which puts it among the top of VC firms investing in China or anywhere. In its latest fund-raising, which includes a third Chinese RMB fund of approximately $225 million, GGV smartly added an Entrepreneurs Fund of $60 million that taps the energy, capital, and knowhow of successful company founders who are among the fund's limited partner backers.

The biggest presence at GGV Capital is Silicon Valley–based

managing partner Hans Tung, a Taiwanese native whom I've known since his days in Shanghai and Menlo Park at Bessemer Venture Partners between 2005 and 2007. His career took off at Qiming Venture from 2007 to 2013, where he championed the firm's investments in Xiaomi, dating from 2010, and made one of his first investments in a Chinese car rental company, eHi Car Services, which attracted a funding partnership with Enterprise and a NYSE listing in 2014. Wanting to move his base to Silicon Valley from Shanghai, Tung left China in 2013, moved to San Francisco and Woodside, and joined GGV Capital. Now he is among the highest-profile venture capitalists in the Valley from cohosting his firm's popular 996 podcast show interviewing tech founders and his partners, sharing views on the firm's WeChat group, judging pitch contests, speaking regularly at tech events, and regularly appearing on CNBC TV as a commentator about China tech. He's made the *Forbes* list of top VCs six times, most recently ranked seventh of the 100 list, and he counts 13 unicorns valued at more than $1 billion in his portfolio, including several in this book: smartphone maker Xiaomi, e-commerce site Xiaohongshu, and music entertainment streamer Musical.y, which was sold to ByteDance for around $900 million in 2017. Tung doesn't mind sharing that an early bet on Xiaomi earned 866 times the initial investment, based on the IPO price in 2018. Since moving back stateside, Tung also spends time working with US companies on China entry plans and has advised Airbnb cofounder Nathan Blecharczyk on the home-sharing company's strategy for China.

He continues to be excited about the China tech opportunity in spite of the overall economic slowdown. "The whole economy is changing its constitution. The new internet-driven economy is upgrading the original offline economy. In that process, the overall growth is slowing but the growth of the net side is growing much faster than the general economy." He pointed to the high growth rates of even established Chinese tech companies such as Alibaba and Tencent as evidence.

Managing partner Jenny Lee also has aced the *Forbes* list. A self-professed geek, she is an electrical engineer by training and a former fighter jet engineer who moved to Shanghai from Singapore and played a major role in setting up GGV Capital in Shanghai. I met her a few years before that, when she was starting to look at China venture while a VP at Japanese investment firm JAFCO Asia. She's famously broken through the glass ceiling of women in venture, on the *Forbes* list of top venture investors since 2012 and crushing it in 2015 within the top 10 ranks and in 2019 in nineteenth place. With her passion for cutting-edge technologies, she's invested in drone startup EHang and its dream of a flying taxi, and in AI language learning bot Liulishuo, also known as LingoChamp, which collected $72 million in an IPO on the NYSE in 2018. Lee was also behind Chinese social media company YY, which has been dominating screens in China with video streaming. YY listed on the Nasdaq in 2012 and has appreciated over 10 times since its public offering.[24]

Damn, That's Good DCM!

DCM Ventures has been at the forefront of Sand Hill Road–anchored venture firms that are bridging East and West and enjoys status for cross-Pacific expertise. The firm has $4 billion under management and has invested in more than 400 companies since its founding in 1996.

I attended DCM's summit at Pebble Beach in celebration of the firm's twentieth anniversary. A highlight of the evening was when DCM cofounder and general partner David Chao chatted with Jerry Yang about the time when the Yahoo! cofounder strolled on that same beach with Jack Ma 12 years ago, a stroll that led Yahoo! to take a 40 percent stake in Alibaba at what was then an astonishing $1 billion.

Certainly, a lot has changed since venture capitalist Chao told me back in 2007 that it would only be 10 to 20 years before China sees the likes of Gates or Jobs. That may have been Jack Ma of Alibaba or Lei Jun of Xiaomi. Yang himself now works with and invests in technology entrepreneurs through AME Cloud Ventures, the innovation investment firm he founded in 2012, and serves on the board of five companies, no longer including Yahoo! but spanning to China-angled companies Alibaba, Lenovo, and Didi.

A differentiator for DCM as a venture firm is its three-market approach, focusing on the United States, China, and Japan with offices in Silicon Valley, Beijing, and Tokyo. China represents one-third of the firm's startup investments, but the majority of its deals are in the United States. The firm's annual holiday party at the San Francisco Museum of Art is a must-attend occasion and attracts many deal makers I know in the cross-border venture business.

The firm's two partners on the *Forbes* list are Chao and Hurst Lin. Cofounder Chao has a track record as one of Asia's top venture capitalists and he's one of the *Forbes* VC hall-of-famers. He's a hard-driving perfectionist, and a long-running joke is that DCM stands for David Chao Management (but actually it stands for Doll, as in legendary VC Dixon Doll, who now has his own firm in the Bay Area, Impact Venture Capital). Since dropping out of medical school and working at Apple in marketing and product management, Chao has had major impact as a cofounder of DCM in 1996. He's also on the advisory board of China's heavyweight Legend Capital, a venture arm of Lenovo his firm has partnered with on deals. Chao has been an investor soul mate to a few tech founders I know—think Joe Chen of social networking site Renren, which scored a NYSE IPO of $740 million in 2001 but who is now tech investing himself after Renren didn't become the China Facebook. Chao has played in all of DCM's geographies and in China counts Nasdaq-listed HR ad service 51job and Wanda-owned 99bill among his major wins.

Another standout partner is DCM China cofounder Lin. He focuses on investments in consumer internet businesses, a sector he knows well from his pioneering days in the development of China's net market as a cofounder of Nasdaq-listed SINA, China's large internet portal. A quick wit, he's found his second calling as a venture capitalist since joining DCM in Beijing in 2006 with a string of successes that regularly puts him on the *Forbes* list. He's well represented in this book with investments in video sharing app Kuaishou and online discount retailer and NYSE-listed Vipshop. Other notable China deals of his are online English-language tutoring service 51Talk, a NYSE-listed company since 2016, and online classifieds website 58.com, trading on Nasdaq since 2013.

Lightning Deals

Not to be overlooked is Lightspeed Venture Partners' affiliate Lightspeed China Partners, led by founding partner and China tech investor James Mi. Lightspeed China has been doing Chinese deals exclusively since breaking off from the mother ship in 2012, setting up in Shanghai and Beijing and racking up 96 investments from seven funds totaling $1.5 billion, including an RMB fund of around $87 million. Kicking off 2019, Lightspeed China pulled in $560 million for two new China funds focused on young and growing technology companies.

Mi joined Lightspeed in 2008 from heading up M&A for Google in China, where he spearheaded investments in Baidu. At Lightspeed China, Mi has been racking up the wins in 2017 and 2018: IPOs of innovative startups Pinduoduo and Metiuan plus internet finance company Rong360 and peer-to-peer lender PPDAI, in addition to a $300 million acquisition of a portfolio company, selfie editing and sharing app FaceU by ByteDance.

"China's enterprise service and deep tech innovation is in the early innings of development. Given China's vast market, deep talent pool, and increasing demand for home-grown deep technologies across various industries, we are seeing accelerated growth and significant investment opportunities," says Mi, who places high in the VC rankings.

On Red

Redpoint China has kept the pace going by launching two new funds in 2019 for early startups and growing companies in consumer, enterprise, and frontier tech sectors. The funds were raised in three months and were oversubscribed, founder and managing partner David Yuan says. The new funds bring the firm's institutional investors past 30, more than doubling the number from an inaugural $180 million fund in 2016.

"The fund-raising climate for China-focused funds is increasingly bifurcating between 'has and has-nots,'" Yuan says. "First-time managers are experiencing increasingly significant challenges in raising money. Many LPs with a China mandate have already reached full allocations, and can only add on new managers when there are slots open."

In 2018, four of Redpoint's Chinese portfolio companies held successful IPOs in the United States, Hong Kong, or China, including gaming company iDreamSky and news aggregator app Qutoutiao.

Gee, GSR!

Last but not least, GSR Ventures factors in for its early start in Chinese venture and its dynamic cofounder Sonny Wu, who would always end interviews with, "The best is yet to come."

He may have been right. Wu now runs mergers and acquisitions fund GSR Global M&A, focused on cross-border buyouts, while GSR Ventures has been a first institutional investor in several new-generation Chinese companies. Beijing-based partner Allen Zhu made it on the *Forbes* list of top 100 venture investors four years in a row, with investments in ride-hailing leader Didi Chuxing and Alibaba-acquired food delivery service Ele.me—but also the doomed bike-sharing startup Ofo.

Center of Gravity

The center of gravity for venture investing has not shifted yet from Silicon Valley to China and it may never. But if and when it does, the group of Sino-US venture capitalists profiled here are the pioneers who were there from the start when tech entrepreneurship and venture capital began to transform China. They bet big on China's fast-growth tech upstarts and didn't miss out. Many of the startups they've invested in have already become stars through high-valued IPOs and acquisitions. The new megafunds they've recently raised will seek out dozens of other catches that could disrupt technologies for China and the world. There's still a long runway. China tech founders are being groomed and capitalized as tomorrow's leaders in a new, vibrant Silicon Dragon tech hub thousands of miles away from the original Valley.

part three

TECH SECTORS THAT MATTER MOST: CHINA'S GRAB FOR SUPERPOWER STATUS

A look at key market sectors with strong potential to shake up leadership and overturn technology standards in the world as East and West vie for tech superpower status.

CHAPTER 6

FACE-OFF IN AI

China and the United States are racing to dominate the high-stakes AI market. China could bypass the United States with its wealth of data and quicker rollout of self-driving vehicles, facial recognition for public security, and AI technologies for fintech, edtech, and health-care startups.

Dr. Xu Li, cofounder and CEO of SenseTime, the world's most valuable artificial intelligence startup, is sharing the story of the company's name in Chinese, Shangtang. The name comes from the phonetic translation of Shang dynasty, an era beginning around 1600 BCE that saw rapid advances in math, astronomy, agriculture, and handcrafts, and from its first emperor, Tang. Presented with an award for Silicon Dragon founder of the year, Dr. Xu told the audience that while he never imagined his startup would become the leading AI unicorn-valued company in the world, his dream from the very beginning was to use AI technologies to have as much impact as advancements in that historic time.

His early investor is famed Hong Kong investment banker Francis Leung, known as the father of the red chips for bringing Chinese companies to list in Hong Kong. Leung nodded in agreement as he sat next to Dr. Xu on stage at the awards program in Hong Kong.

Leung said he believes SenseTime will be in the class of Baidu, Alibaba, and Tencent, at the top of China's AI sector.

SenseTime has perfected camera surveillance technology that analyzes faces, car license plates, vehicle types, and events for public security in China. Its high-tech system also verifies identities for payments at staff-less checkouts, peer-to-peer lending, and phone unlocks. During Chinese New Year, when travel is at a peak, SenseTime facial recognition technology lets passengers match personal ID cards with their tickets and luggage and cut waiting time in long lines. They also can check boarding times, flight status, and gates by standing in front of a "smart" sensor camera. This technology is being rolled out at Beijing's new airport, but don't count on it showing up at O'Hare Airport anytime soon.

In China, cameras are recording footage at major intersections and in public places, 200 million of them across the country that can catch jaywalkers, break-ins, and shoplifting. China's surveillance state assigns a score to every individual based on their social behavior, which can impact their credit rating and even their ability to buy train tickets. This no doubt conjures up images of Big Brother watching all that goes on, not to mention an episode of *Black Mirror*. But it's not just China. The New York City Police Department is reportedly monitoring citizens using cameras and facial recognition software developed in China, from SenseTime partner Hikvision.[1]

In the United States, tech giants Google, Microsoft, Amazon, Facebook, and IBM dominate AI for many futuristic and practical uses. Google self-driving cars are being tested on California's Highway 101; Facebook spins out posts based on deep learning of content preferences; Amazon's Alexa powers lights, TVs, and speakers by voice activation; and Microsoft's Azure relies on cognitive computing for speech and language applications, while IBM Watson's AI-based computer system increases productivity and improves customer service for call centers, production lines, and warehouses.

In China, Baidu, Alibaba, and Tencent are working on similar technologies and racing with the US tech giants to become world leaders in AI. The Ministry of Science and Technology in China has earmarked specialties for each of these Chinese tech titans in its master plan for AI global dominance: Baidu for autonomous driving, Alibaba for smart-city initiatives, and Tencent for computer vision in medical diagnoses. The Chinese government also has designated two startups to lead AI development: SenseTime for facial recognition and iFlytek for speech recognition.

Baidu, Alibaba, and Tencent are all powering up in autonomous driving, and each has a specialty focus area in AI. Baidu has its DuerOS line of smart household goods and Apollo, an open platform for self-driving technology solutions, and detoured on the AI journey several years before Google in 2015. Alipay uses facial recognition for payments, and Alibaba has an AI cloud platform called City Brain that crunches data and determines patterns for better urban planning. Tencent is integrating rich media formats such as face-swapping effects and video chat filters into its social media and is investing in personalized medicine, digitized patient health-care records, and remote health-care monitoring. In their quest to win the AI challenge, the three titans are hunting for new AI technologies and applications by investing in AI startups globally. Since 2014, this trio of Chinese tech giants has made 39 equity deals in startups that are building AI chips and software.[2]

Despite scrutiny over Chinese investments in US technology startups, this cross-border pipeline is active in artificial intelligence. In the United States, Tencent has made the most deals but Baidu has the most diversified AI portfolio, spanning to health care, advertising, and media startups. Baidu's AI plate includes not only 95 partners in its ecosystem worldwide working on autonomous driving but also investments in AI-related startups in the US: ZestFinance in fintech underwriting, Kitt.ai in conversational language search, TigerGraph in data link analytics, Tiger Computing Solutions in big data, and xPerception in computer vision for self-driving. Tencent has a number of AI partnerships in the health-care space globally and has invested in 12 US startups in AI, including avatar creator ObEN and two in drug discovery based on deep learning, Atomwise and XtalPi.

White House Weighs In

China hasn't created any world leaders in cars or semiconductors, but few pooh-pooh its growing ability of AI fundamental technology that touches our everyday lives, from e-commerce fraud detection to systems that can detect cancer; to sensors for self-driving; to robot-powered deliveries, education, and online lending.

The advent of AI has both US and Chinese tech companies pouring money and talent into AI technologies and applications. At stake is a possible shift of the world order in the global economy. China is pushing harder and faster to be the global leader, to create a $150 billion industry by 2030. The United States has long held the lead in AI talent and research, but China is gaining on the United States for the number of high-impact scientific papers published on AI.

The importance of AI and the race for global tech leadership has gotten special attention from the White House as frictions have mounted over concerns that the United States is not keeping pace with China. In February 2019, President Trump launched what he termed an "American AI Initiative" to direct federal agencies to prioritize investments and R&D to accelerate its adoption—though the directive lacked increased funding. This US national strategy for AI follows China's plan, unveiled in 2017, to become the world leader in AI by 2030, with two Chinese cities set to invest $7 billion in the effort.

The future of AI development is far-reaching and deep, from increased productivity and labor automation to advances in smart health care. AI advances could lead global GDP to grow by 16 percent to $13 trillion by 2030, according to McKinsey Global Institute, equivalent in impact to the steam engines of the 1800s, robots in 1990s, and the internet since 2000. A separate study by PricewaterhouseCoopers China predicts AI will boost GDP 26 percent by 2030 to $15.7 trillion and additionally notes that the greatest economic gains from AI will go to China (26 percent boost) over North America (14.5 percent gain).[3] One advantage China has in the AI race is that it does not have the same restrictive privacy laws as the United States, making it easier to collect large data sets, which are then used in recognizing patterns for machine learning, a subset of artificial intelligence.

AI expert and venture investor Kai-Fu Lee underscores that the United States has long been, and remains, the global leader in AI for research and hardware. But he emphasizes that China is catching up at an astonishingly rapid rate in implementing the technology in practical ways. China has an advantage based on large numbers of well-trained AI talent, a supportive government policy, and access to a vast amount of data sets powering AI and gleaned from China's world-leading number of internet and mobile phone users, he notes. In the age of AI, data is the new oil, so China is the new Saudi Arabia, says Lee, author of *AI Superpowers*.[4]

His venture investment firm in Beijing, Sinovation Ventures, which I've visited multiple times, is betting on AI's future. Lee, who is widely known for his pioneering work in speech recognition and artificial intelligence, is an investor in five Chinese AI companies worth more than $1 billion. Two that are in the forefront are Megvii, a Chinese developer of facial recognition system Face++, and 4Paradigm, a machine learning software for detecting fraud in insurance and banking. I've known Lee since 2006, when he was running Google China, and I've watched his career flourish as a China tech investor from starting Sinovation Ventures in

1999 and as a world-leading AI expert. Anita Huang, his operating partner in Beijing whom I've known since she worked on building up YouTube-like Tudou in the early Chinese internet era, took me around the venture firm's showroom featuring its cutting-edge portfolio companies. One standout was an AI-integrated interactive window that is used by Withwheat, a high-end bakery chain in Beijing, to run no-cashier outlets.

The search is on to develop and implement more AI technologies, and Lee's firm is taking a lead. In 2018, Sinovation Ventures raised a $500 million fund and a Chinese currency fund worth $375 million to further invest in AI startups in China. The firm also is running an incubator that researches next-generation technologies for AI and then funds and works with founders to develop them. It's opened a school for AI too, with the support of China's Ministry of Education and Peking University, to train top engineering and science students in machine learning and other AI techniques. Sinovation Ventures additionally is a coorganizer of an AI contest in China with a sizable $450,000 prize.

The transformative AI technology field has emerged as one of the hottest sectors for tech investment in startups. Both of the world's AI superpowers are racing after the opportunity, with heavy investments by China and the United States and China in particular gaining momentum. As China moves toward an AI-based economy, its venture capital investments in AI have stacked up. China weighed in with 48 percent of $4.9 billion in funding and 10 percent of deals in 2017, led by several megarounds for Chinese AI startups, and for the first time surpassed the United States at $4.4 billion. In 2016, China accounted for only about 11 percent of global funding.[5] The United States is not about to lose advantage. Venture capital investments in AI startups in the United States have increased significantly every year since 2013 and jumped sharply by 72 percent to $9.3 billion in 466 startups in 2018, though the number of deals dipped from 533 the previous year.[6]

SenseTime Leads Unicorn Pack

The global AI unicorn club is getting increasingly crowded—17 new AI startups earned that status in 2018, up from 11 the year before. Of 32 AI unicorns worldwide, China counts nine. SenseTime leads the pack as the world's highest-valued unicorn at $4.5 billion and takes special credit for being invested in by both Alibaba and Tencent. SenseTime competes most closely with Megvii and its visual computing and facial recognition Face++ platform, which is valued at $1 billion from a $608 million funding in October 2017 led by Alibaba's Ant Financial Services along with Sinovation Ventures.

Since its start in 2014, SenseTime has been racking up the kudos. It boasts more than 500 patent filings worldwide, some 700 corporate customers and partners in China and overseas, and market leadership across several vertical sectors: smart city, finance, retail, education, and real estate. The Chinese AI startup reached profitability in 2017, and its growth rate is off the charts, soaring by 400 percent year-over-year in a recent three-year period.

SenseTime is no slacker either when it comes to funding. The startup has collected $1.6 billion in venture capital, most recently $620 million in May 2018 at a valuation of more than $4.5 billion from several prominent firms including Silver Lake, Fidelity International, and Tiger Global. Just two months before, SenseTime pulled in $600 million led by Alibaba—its largest shareholder—with Singapore's state investment firm, Temasek. The year before, SenseTime reeled in $410 million led by private equity firm CDH Investments and earlier backer IDG Capital. The huge SoftBank Vision Fund could be next to fund SenseTime.

University Spinout Does Good

SenseTime grew out of the computer vision lab at the Chinese University of Hong Kong, led by Professor Tang Xiaoou along with a group of students who developed facial recognition algorithms that had higher accuracy rates than human eyes. SenseTime Group was spun off as a business while housed at the Hong Kong government–backed, campus-like Hong Kong Science and Technology Park. One of Tang's protégés, Dr. Xu, a 2010 PhD graduate in computer science at the Chinese University of Hong Kong, was made CEO after years of being in the research trenches. SenseTime began launching products in 2016 with Xu, 36, at the helm.

By 2017, SenseTime grew to 1,000 employees, including 140 PhDs drawn from local universities as well as globally elite Stanford University, Tsinghua University, and the Massachusetts Institute of Technology. SenseTime has scored commercial success with the support of the Chinese government, which has provided data sets from various agencies to train the company's algorithms to better reference graphics, images, and objects much faster and far more precisely than the human brain. In China, about one-third of its customers are in the public security sector, including police bureaus in southern China that use the system to identify criminals. The client list also includes Chinese smartphone makers Xiaomi and OPPO, social networking service Weibo, Hainan Airlines, and payment systems China UnionPay.

Powering Up with Honda and Qualcomm

Outside of Mainland China, SenseTime has established a subsidiary in Japan to work with Honda to develop autonomous driving and has unveiled its own dedicated testing course for self-driving vehicles near Tokyo. It's also linked up with US chip maker Qualcomm, a strategic investor in SenseTime, to integrate the Chinese startup's proprietary algorithms and image recognition capabilities into next-generation mobile devices. And SenseTime has joined an alliance led by the Massachusetts Institute of Technology to advance AI across a number of fields.

Pride in the startup's accomplishments and China's advances

are certainly understandable. "The world is facing east," said Xu at a recent China entrepreneurial summit. In the past, "We journeyed to the West because we relied on the West for advanced technology and industrial modes. Now we must embark on a treasure voyage to better service other industries with advanced technology."[7]

AI at Work in Edtech: LAIX

Chinese startups are rapidly implementing AI technology into their businesses to gain an edge. Several startups I know are leveraging AI across diverse fields: LAIX in edtech, WeLab in fintech, and Link-Doc in medical diagnoses.

Leading the way in a wave of AI-linked startups expected to go public soon is LAIX, a Shanghai-based developer of an AI-driven English tutoring system. Known in China as Liulishuo, LAIX went public on the New York Stock Exchange in October 2018 and raised $71.9 million. The innovative company got a head start with an expert team and a core group of plugged-in investors, among them GGV Capital, Hearst Ventures, and Cherubic Ventures. The AI-based edtech startup was founded in 2012 by Wang Yi, a Princeton PhD in computer science and former Google product manager in charge of analytics and cloud computing. Yi's startup is disrupting the online education sector by helping Chinese people learn to speak English through AI-powered interactive, customized courses accessed on mobile phones. Its AI technology crunches data to feed a speech recognition engine that can provide feedback on pronunciation, grammar, and vocabulary. This being China, LAIX integrates games and social sharing into its mobile app to make for a more fun, interactive learning experience. Riding high on China's growing trend toward online knowledge platforms, LAIX attracted 110 million registered users in 2018, including 2.5 million who paid for courses for the full year. Revenues soared in its first year as a publicly traded company by 285 percent to $93 million.

AI at Work in Fintech: WeLab

An example of AI disrupting traditional banking comes from Hong Kong–based fintech startup WeLab, which provides small consumer loans in an online instant, with fewer than average defaults by relying on AI and data to determine creditworthiness. WeLab technology combs through online data such as bill payment records and social media profiles to figure out which potential borrowers are likely to pay their loans on time. Then it prices and tailors online consumer loans. Consumers complete the entire lending process over their smartphone and don't need an established credit history—an issue among young people starting in their careers. Loan decisions for individual borrowers are made online within seconds. One hint: don't fill out the online form in all capital letters. WeLab has found applicants who write in upper case are not good credit risks.

A technology team of more than 210 engineers and data scientists have guided WeLab in reinventing traditional lending and assessing credit risks by three proprietary AI systems: WeDefend detects fraud and suspicious behavior by analyzing more than 2,500 user data points in under one second. WeReach peeks into consumers' influence and interactions with social connections. WeFlex monitors changes in consumer behavior that could signal trouble with collections so that credit limits or terms can be changed.

WeLab charges relatively high interest rates of 25 percent, and its delinquency rates are small—only 1.5 percent in Mainland China and much less than that in Hong Kong.

Riding on China's advanced fintech and mobile payments market and reinventing the cumbersome loans process, WeLab has attracted over 32 million users and grown into a profitable $155 million business. WeLab was gearing up to go public in Hong Kong in 2018 but has since postponed that move. The online loan startup has recently been awarded Hong Kong's fourth online-only banking license, which should escalate its business and could lead to a rethink of that IPO. The firm is a few years ahead of its closest US comparable, LendingClub, in applying AI to lending. In China, WeLab has multiple competitors: online banks from Chinese tech giants Ant Financial and Tencent WeBank in addition to Chinese peer-to-peer lending services, among them two New York–traded companies, Alibaba-backed Qudian and Legend Capital–invested PPDAI Group. The peer-to-peer lending sector has recently cooled after a regulatory crackdown.

Stanford Founder

The brainchild of WeLab is CEO, chairman, and founder Simon Loong, a Stanford University business school graduate with a 20-year banking career in Asia at Citibank and Standard Chartered Bank. He launched WeLab in July 2013 as Hong Kong's first online lending platform and expanded a year later to Mainland China

Founder Loong has used his connections and knowhow to bring in a well-credentialed group of investors that have put WeLab in the unicorn-valued league: initially $20 million in January 2015 from Sequoia Capital, DST Global founder Yuri Milner, and the technology investment unit TOM Group of tycoon Li Ka-shing's conglomerate CK Hutchison, and a year later $160 million from a consortium of investors led by Malaysian sovereign wealth fund Khazanah Nasional Berhad and including Dutch multinational ING Group. Another $220 million came in from Alibaba Entrepreneurs Fund, the International Finance Corporation, Credit Suisse, and China Construction Bank. If and when WeLab goes public, they stand to benefit.

AI at Work in Health Care: LinkDoc

Another good example of AI at work comes from fast-growing medical AI startup LinkDoc. The young company, based in Beijing, specializes in handling cancer diagnoses. It's well positioned in China's growing aged population and in a sector that the government has earmarked as part of its Made in China 2025 initiative.

LinkDoc relies on machine learning tools and an image diagnosis system that can reduce misdiagnosis and inaccuracies. In just three years, the AI health-care startup signed up 1,000 hospitals in more than 30 provinces in China and accumulated data on 3 million cancer patients in China. It's adding data on 200,000 patients per month in China, which has the highest cancer rates in the world.

> *"This would not have been possible before. I would never have foreseen this."*
>
> **Xiaodong Jiang**
> Managing partner, Long Hill Capital

Founded in 2014, LinkDoc was first invested in by Chinese venture firm Long Hill Capital in 2015, and many other investors have followed. The startup's fourth round of funding, at $151 million in 2018, pulled in two sovereign wealth funds, Temasek and China Investment Corp.

LinkDoc reached $70 million in revenue its first year—a solid result early in the life of the biotech firm. The growth was made possible by the integration of AI and big data technology into biomedicine to develop oncology media records covering all types of cancers. "This would not have been possible before" in the health-care field, said Xiadong Jiang, managing partner at Long Hill Capital. "I would never have foreseen this."

CHAPTER 7

A SHARED ECONOMY

The fad in China's booming sharing economy for hitching a ride has been very bumpy for bike-sharing startup Ofo, but ride-hailing leader Didi has had a good run in beating Uber. Now shared umbrellas, mobile chargers, and even takeout kitchens are here.

An hour's ride from Beijing's Forbidden City to the high-tech zone Zhongguancun in the northwest of this sprawling megalopolis is the modern headquarters of Didi Chuxing, China's ride-hailing service, which ranks among the most valuable venture-backed startups worldwide. You know you've arrived by the colorful Didi taxi sculpture parked out front. Didi is positioned in China's sharing economy sector, the nation's leader in the world's largest ride-hailing market, worth $30 billion. The privately held startup, funded by Tencent, Alibaba, and even Apple, is perhaps best known to those outside of China as the company that won the competition with Uber in China in a fierce three-year battle that ended in 2016 when Uber sold out to its China rival.

By absorbing Uber and its main Chinese rival, Kuaidi, Didi climbed to the top of the ride-hailing heap with a dominant share of China's large $23 billion ride share market, which is projected to nearly triple by 2020.[1] On the downside, Didi has recently undergone layoffs of 15 percent of its workforce in China's economic slowdown, faced closer government scrutiny over safety issues, and failed to turn a profit since starting in 2012. An IPO that was anticipated a few years ago has been put on hold.

A transition to new technologies and rival entries from Tencent and Alibaba require Didi to invest more and more to stay even. Didi has raised $21 billion in 17 rounds of funding.

Bike-Sharing Pedals Away: Ofo and Mobike

As has been seen before in other instances, it can be difficult to sustain a winning streak in China's topsy-turvy sharing economy—same as in the United States but at a steeper level. Bike-sharing startups Mobike and Ofo came onto the scene a few years ago as a new model of dockless shared bicycles and took off almost overnight as a superfad. City streets and sidewalks were clogged with these colorful bikes. Flooded with venture money and backing from Alibaba and Tencent, startups quickly overexpanded in China and overseas, chasing the opportunities with heavy subsidies for riders in a fiercely competitive and opportune, but ultimately money-losing, market. Several bike-sharing startups flopped and went out of business. The bike-sharing era has gone from bunches of colorful, shiny bikes parading city streets and sidewalks to rusted-out frames ditched along the road. Fads come and go quickly in China. Then again, my neighborhood south of San Francisco had lime-colored bike-sharing rentals for about one year. But then, the well-funded startup that was providing them for rent, Lime-Bike (renamed Lime), suddenly collected all the bikes one weekend and since has tried unsuccessfully to get the city to switch to scooters, which could be a more profitable exercise for the young business.

As much as $2.2 billion of venture money was poured into China's pioneering bike-sharing player Ofo, led by Alibaba as the biggest investor along with experienced venture capital firms Matrix Partners China, GSR Ventures, ZhenFund, and Yuri Milner's DST Global. Now they are on the hook as Ofo has withdrawn from overseas markets and seeks to reboot its money-losing business in China and move toward an e-commerce model converting the customer deposits it's collected for discounted items. A more fortunate scenario has occurred for Mobike, which was acquired by food services app Meituan for $2.7 billion in 2018 after being pumped up with investments of $900 million from Tencent and leading China-side venture investors Sequoia Capital, Qiming Venture, and Hillhouse Capital. The Meituan management team is gone, and so is the Mobike brand name, renamed Meituan Bike as its new owner seeks to make the operation more efficient and trim substantial losses.

Kitchens on the Go: Panda Selected

Another sharing economy concept that originated in China is shared kitchens, thanks to Chinese startup Panda Selected. The sharing is mainly for takeout restaurants and catering operations, not your home, and meant to reduce their overhead costs and meet demand 24 hours. Panda Selected is disrupting the traditional restaurant model, similar to how bikes, food delivery, hospitality, and office spaces have undergone change—and yes, ride-hailing services too.

The Beijing-based startup, founded by CEO Li Haipeng, has recently attracted $50 million in funding from DCM, Tiger Global, and others, taking its total funding to $80 million. Within the three years of its start in 2016, Panda Selected has grown to 120 locations in China's major business hubs. This service is meant to attract a young on-the-go population who order food by mobile app. Former Uber CEO Travis Kalanick is working on a similar idea with his Los Angeles–based startup CloudKitchens, so perhaps this innovative concept will become better known in the United States.

The sharing economy has arisen in China with the uptake of mobile apps and payments and a young consumer population that enjoys experimenting with new things. The appeal of ride hailing is the ability to tap on a mobile screen and secure a driver to take you where you want to go for less than a taxi fare, then step out of the car without dealing with cash.

Didi has proven to be an innovator in ride hailing, a segment that has gotten a lot of attention with the recent public offerings of Uber and Lyft in the United States. One Didi service sends a driver to your personal car when you've had too much to drink. Another is an SOS feature to activate in case of a hazard or emergency.

Today, in China's congested cities, it's no longer a status symbol to own a car. It's a pain because of traffic jams, parking hassles, and financial costs. China has more than 300 million drivers, but only about 20 percent of China's 1.4 billion population own cars. Car rentals are not popular. This creates a big opportunity. About two-thirds of Chinese consumers have tried out the service, compared with only about one-fourth in the United States.[2]

Didi: Ride-Sharing King of the East

Quite appropriately, the name Didi means "beep-beep" in Mandarin. Mobility should be its middle name. Didi serves 500 million individuals with taxis, buses, and bikes; offers flexible work to 31 million drivers; and facilitates 30 million rides each day.[3]

Table 7-1

At a Glance: Didi

Location: Beijing

Founder: Cheng Wei

Launch: 2012

Status: privately held

Financials: estimated revenues of $1 billion, unprofitable

Notable: beat Uber in China

A walk through the exhibit hall at Didi's headquarters paints an overall picture of its technologies and business development starting as a taxi service. The corporate public relations master who speaks fluent English points to this piece of trivia to illustrate the company's fast growth: on one snowy day in Beijing in November 2012, daily trips on Didi exceeded 1,000, while 74.3 billion passenger trips were made in 2017.

Didi's continued dominance is not guaranteed. New entrants in China such as WeChat and Alipay are competing and starting to make inroads—half of all Didi rides are ordered from WeChat and Alipay apps, compared with 40 percent through the Didi app.[4] Autonomous driving is coming for taxis, and auto manufacturers have jumped into ride hailing. At the start of 2019, Didi was reorganizing, laying off 2,000 staff, or 15 percent, of its workforce and stepping up investment in geographic expansion, driver training, and safety measures.

The most troubling issue for Didi is passenger safety, a problem throughout this market sector internationally. After the murder of two female passengers in China, Didi in mid-2018 suspended its carpooling service, operated by private drivers. Downloads for the Didi app plummeted after the incidents. Didi has been dealing with the crisis by introducing several safety measures in China that include verifying its drivers with facial recognition tests, installing emergency buttons for both drivers and passengers, and such extreme measures as using the driver's phone to audio record trips—with the passenger's consent—that are stored and then deleted at Didi within one week. Not sure if Uber will be trying this out in the United States.

The Traffic Brain

In some other realms, Didi sees a brighter horizon. The company is focusing on expanding outside China, investing more in AI systems and autonomous driving, conducting research at a Silicon Valley lab, and planning an electric vehicle network of 10 million by 2028. Like Uber and Lyft experimenting with new self-driving thrills, Didi is testing self-driving vehicles in four cities in China and the United States and has a grand plan to launch driver-less taxis soon. Robo taxis are already a reality in China—and the United States. The self-driving highway is looking more and more jammed. Pony.ai, a Chinese autonomous car startup, recently launched a test of a self-driving taxi while Waymo, the spinout from Google's self-driving research, is testing a service in Arizona and Uber has restarted tests of its service after crashes in initial 2018 trials in Pittsburgh and Arizona.

At Didi, I met with Tiger Qie, vice president and general manager of marketplace and apps. In this role, he oversees driver dispatching, carpooling, supply-and-demand forecasts, incentives, route planning, navigation, and mapping technologies. It's a lot to handle at a huge operation like Didi, and Qie's schedule is packed.

From him, I learned about the Didi Brain, an AI system that leverages big data analysis, machine learning, and cloud computing technologies to maximize the transit system's efficiency, much like Uber and Lyft do. The Didi Brain uses algorithms to forecast passenger demand and driver supply in various areas at any given time to offer the shortest routes, alleviate traffic congestion in cities, and match riders with drivers in single rides or pools. Massive amounts of real-time data estimate arrival times more accurately than traditional computing methods. Didi drivers are rated based on service scores from customers. Drivers with high scores earn more. Order cancellations are handled by an AI-powered system that takes 10 milliseconds to decide if the driver or rider is responsible for a cancellation. Another system, Didi's so-called Traffic Brain, integrates data from roads and maps, analyzes it, and then controls traffic within cities by timing traffic signals and adjusting reversible lanes. Didi has launched these smart-transport technologies in 20 Chinese cities to help control traffic flow and ease congestion.

What else does this transport brainpower do? Didi has developed a global R&D network of big data and intelligent-driving technologies and opened a research lab in Mountain View and an AI lab in Beijing. The Chinese company also has formed innovation partnerships with Stanford University and the University of Michigan to accelerate development of smart transport.

Didi vs. Uber

It's interesting to compare and contrast Didi in China with Uber in the United States. They are more alike than you might imagine. See table 7-2.

Table 7-2

Comparing Didi and Uber

Start Date and Headquarters

Didi: 2012 in Beijing

Uber: 2009 in San Francisco

Venture Capital Raised

Didi: $21 billion

Uber: $20 billion

Number of Users

Didi: 40 million monthly users

Uber: 40 million monthly users

Locations

Didi: 400 cities

Uber: 400 cities

Market Share in Home Market

Didi: 80 percent

Uber: 73 percent

Backers

Didi: Tencent, Alibaba, SoftBank, Apple, and Singapore government investment fund Temasek

Uber: SoftBank, Baidu, Google Ventures, private equity firm TPG, and Silicon Valley VC firm Benchmark Capital

Market Valuation

Didi: $4 billion financing in late 2017 from SoftBank and an Abu Dhabi state fund with $56 billion valuation, ranked third among global unicorns

Uber: investment in early 2018 from SoftBank consortium with $72 billion valuation, ranked second among global unicorns

Public Listing Plans

Didi: plans to go public have been delayed and company is restructuring

Uber: went public in 2019

Didi Buys Out Uber in China

Contrasts between Uber and Didi take on a whole new meaning when you consider what happened to Uber in China. Uber founder

Travis Kalanick battled against two Chinese executives at Didi: founder and CEO Cheng Wei, a former sales manager at Alipay, and Liu Qing (aka Jean Liu), a sophisticated and polished Goldman Sachs alumnus from Hong Kong with a Western-style PR-ish manner.

Kalanick took Uber to China in late 2013 and spent heavily to gain market share by subsidizing rides and driver salaries. He was widely quoted as being gung ho about winning the China market and joked that he was spending so much time in the country that he should apply for Chinese citizenship.

Kalanick led a three-year, all-out effort to win the Chinese market. He raised more than $1 billion from Chinese investors and grew the team to 800 in 60 cities. He partnered with Baidu for mapping navigation and with Alipay for payments. He heavily subsidized rides to retain drivers and riders. He dealt with local and national government regulations to standardize the ride-hailing business. But Uber China was burning enormous sums to chase market-share leader Didi Chuxing. The Uber app was considered not as user friendly as the Didi one, lacking some small features such as a way to charge tolls in China.

One thing Kalanick didn't foresee was that his two largest Chinese rivals—Didi and Kuaidi—would merge. In China's cut-throat and fast-paced market, mergers can be a very quick way to get bigger and possibly better. Uber's Chinese competitors joined together in February 2015 and rebranded as Didi Kuaidi. The consolidated company claimed an 80 percent market share of China's ride-hailing market—comparable to Uber's dominant share of the US ride-hailing sector.

Kalanick wasn't about to give up. He took his venture capitalist's advice: be respectful to China officials, call on mayors, and take note of their key performance indicators to win them over. He finally gave up in China after bumping up against new national and provincial regulations for ride-sharing services in 2016—and getting an offer he couldn't refuse.

Faced with some $2 billion of losses in its Chinese operations and the prospect of investing billions more, Uber's China business was bought out in August 2016 by the homegrown champion Didi in a huge $35 billion deal. The transaction did not leave Uber and its investors empty-handed—they got around a 20 percent share of the Chinese company and potential upside from its new owner's growth in China. Uber stopped the heavy subsidies to retain drivers and riders. In an email to the Uber China team when the deal came through, Kalanick called the effort big and bold, particularly for a relatively small startup, and noted that China was uncharted territory. To Carmen Chang, who heads up Asia investments for venture firm NEA, an investor in Uber, this counts as a win. Didi founder Cheng got a board seat at Uber while Kalanick became a board observer at Didi.

The Didi deal wrapped in two Chinese tech titans: Tencent was a backer of Didi while Alibaba funded Kuaidi. Apple hitched a ride too, injecting $1 billion in the newly merged Didi business in May 2016 and getting a board seat in what was seen as a *guanxi* or relationship-building move in China.

Next Move: International

Having won the skirmish in China against Uber, Didi is focusing on thinking and acting globally, offering an English-language interface and accepting international credit cards. Expanding outside China is eating into its coffers, but it's a strategic priority given the possibility of more regulatory crackdowns in China.

Spanning outward, Didi has recently started trial operations in two Australian cities, introduced taxi and mobility services in northern Taiwan, and launched a joint venture in Japan with Soft-Bank to offer taxi-hailing services in Osaka and expand the network to Tokyo and several other Japanese cities. Didi also has upgraded its taxi-hailing services in Hong Kong and begun offering digital payment options including international credit cards, WeChat

Pay, and Alipay. Latin America is not too far away for Didi. The Chinese company entered the Mexican market with a pilot operation in Toluca, the capital of Mexico's central state of Mexico, and launched services in large cities Guadalajara and Monterrey. In Brazil, Didi made a $100 million strategic investment in 2017 in Uber rival startup 99 and then acquired this Brazilian business in 2018 at a valuation of $1 billion, taking direct aim at Uber in Latin America and its two busiest cities, Rio de Janeiro and São Paulo. In announcing the Brazilian deal, Didi CEO Cheng noted the obvious: globalization is a "top strategic priority."[5]

Reaching deeper into Southeast Asia, Didi has funded regional ride-hailing leaders Grab in Singapore and Ola in India. These southeastern startups are also invested in by Japanese tech conglomerate SoftBank, which has strategically put money behind one ride-hailing entrant in each region, including Didi in China.

In a repeat of Uber's saga in China, regional leader Grab—backed by Didi, SoftBank, and Alibaba—acquired Uber's Southeast Asian business in 2018, then merged it.

Don't look for Didi to try entering the United States and compete with Uber on its home turf. Uber and Lyft are already too well entrenched, and battles for position have intensified with Lyft claiming a 35 market share next to dominant Uber, which faced several troubling scandals. With both now publicly traded companies, they could spark a sharing economy IPO parade. But there's a lot of headway to be made.

When Will Didi Make Money?

Getting to profitability has remained a struggle for the privately held Didi, as with many fast-growth tech companies in China. Didi cut back on subsidizing drivers and passengers on a large scale after the Uber battle ended, but it was still losing hundreds of millions of dollars, caught up in a cash-burn spiral of subsidies and discounts.[6] The company's goal of turning a profit in 2018 on net revenues of close to $1 billion[7] evaporated as losses reached $1.6 billion that year. Going public soon looked more and more remote for Didi.

Don't Hitch a Ride

It's been tough overcoming the public image problems and dealing with how to fix serious safety issues. Soon after the two tragic incidents with female passengers, Didi shut down its money-making Hitch pooled-ride service, which had attracted lots of customers with less expensive fares but, unfortunately, some shady drivers. Didi was blamed for using suggestive ads promoting the pooled Hitch service as a way for drivers to meet women and allowing drivers to exchange comments about passengers' looks. Didi issued a public apology, fired two executives, and worked on safety improvements. Meanwhile, the Chinese government banned such carpooling services until tighter safety measures could be put in place.

Let me put in an aside here: I avoid riding alone or at night in a taxi or car service in China since having my own safety issue when a driver in Beijing grabbed at me in the back seat. I find there's safety in numbers on Beijing's subway, and, besides, the subway costs less than $1 and doesn't get stuck in traffic jams.

In addressing the issues, Didi management glumly acknowledged that ignorance, pride, and a superheated market race caused them to lose their way. In a joint statement responding to the crisis, top executives Cheng and Liu said Didi will now "prioritize safety as the most important performance indicator. We raced nonstop, riding on the force of breathless expansion and capital through these few years; but this has no meaning in such a tragic loss of life."[8]

There's a lesson here for managers of other highly competitive, hard-charging tech companies, whether in China or not. But in China right now, the stakes are higher and the battles more intense.

CHAPTER 8

E-COMMERCE
GETS SOCIAL

Just when you think there's nothing else new in e-commerce, along comes social commerce app Pinduoduo, which combines bargain shopping, games, and social sharing—and it's very popular in China's rural areas.

Colin Huang started a social commerce app in China when the e-commerce landscape was already saturated. But within just three years of the app's start in 2015, Pinduoduo went from zero to $278 million in revenues, attracted 300 million users and 1 million merchants, sold $21 billion worth of merchandise, and fetched a market valuation of nearly $24 billion from an oversubscribed $1.7 billion Nasdaq IPO in July 2018. This is quite the punch in China tech power.

Pinduoduo has carved out a niche in two ways. The mobile app combines bargain shopping, gaming, and social media—the drivers of e-commerce today in China. And it's most popular in China's rural areas and lower-end demographics—a new market that has sprung up as the internet has spread outside Beijing and Shanghai.

Pinduoduo has surged to become China's third-largest shopping site, right behind Alibaba's Taobao and JD.com, challenging these 15-year-old e-commerce giants in the country's hypercompetitive market.

"Everybody thinks that China's BAT are so big that it would be impossible to disrupt them, but there's always a company in the garage that is going to come out and do something different," says James Mi, founding partner of Lightspeed China Partners, who invested $10 million in Pinduoduo at its start.

Founder Huang, a Chinese serial entrepreneur with three prior startups and Google experience, describes his startup as a combination of discount retailer Costco and entertainment property Disneyland. But Pinduoduo, which loosely translates as "much more together," is more like Groupon with games and social networking mixed in. Unlike Groupon's group-buying model for discounts that focus on coupons for restaurants, travel, and attractions, the Chinese app is a giant flea market on a mobile platform. Bargain hunters select an item they want to purchase and then invite friends, family, and social connections to click and buy online with them. Prices are slashed with each new buyer in the team purchase. A dashboard keeps score of which users made the most money by inviting friends to join in on the mobile shopping expedition. As tech investor and podcast cohost Rui Ma explains, it can be difficult for friends to turn down an offer to help them save money. "As I think about psychology, it's really ingenious," she says.[1]

Gaming-type promotions such as one-hour specials, cash rewards for daily check-ins, price cuts, lotteries, and lucky draws are blended in to encourage impulse shopping and make it entertaining and fun. Using AI algorithms based on data gathered from customers, Pinduoduo predicts shopper preferences and recommends goods.

"Can we say Pinduoduo out-Grouponed Groupon?" writes China social marketing expert Thomas Graziani, adding that "the shop-with-friends app combines a group-buying strategy with cheap products and social media."[2]

Best sellers are common goods like umbrellas, laundry detergent, tissue paper, and lemon tea. Prices are lower than on Alibaba's

Taobao, even for the same SKUs. The average item sells for $6. The group bargain shopping app has its strongest appeal among China's sizable population of lower-income residents in smaller cities, and the key demographics are young folks with time on their hands and middle-aged housewives. The

> *"Everybody thinks that China's BAT are so big that it would be impossible to disrupt them, but there's always a company in the garage that is going to come out and do something different."*
>
> **James Mi**
> —Founding partner, Lightspeed China Partners

older-guard e-commerce companies JD.com and Alibaba dismiss Pinduoduo and doubt it will survive long. But the fast uptake put Alibaba's Taobao on the defensive.

Pinduoduo's business model is built on driving bulk sales with low-ish prices. Factories and merchants use the app to get rid of low-value products that are overstocked. Pinduoduo's timing was good. Its app has scaled fast by leveraging China's widespread mobile use and by linking up with the popular social network WeChat. A significant amount of traffic for the social commerce app comes from Tencent's WeChat, which is where the teams form to shop. Pinduoduo latches on to one-click WeChat payments and automatic billing to WeChat accounts.

Pinduoduo makes money primarily from collecting fees for marketing services such as pay-for-search keywords and ad placement, and most of the rest comes from charging vendors a small 0.6 percent commission.

Like other fast-growth Chinese startups, the social shopping upstart Pinduoduo has incurred net losses since its start, and that's sparked eagle-eyed investors to think twice.[3] Pinduoduo has a fundamentally flawed business model that "feeds off the scraps of China's retail commerce market, sells the cheapest goods at bargain-basement prices to the poorest people in China's poorest cities, and it's losing money doing this," says hedge fund investor Soren Aandahl at Blue Orca Capital.[4]

Pinduoduo also has been hit with complaints about merchants selling poor-quality items and fakes of popular consumer brands ranging from Louis Vuitton bags to TVs to Pampers diapers. Product quality can be an issue, causing customers to leave after getting fed up with cheaply made products. There have been reports of electric toothbrushes that don't turn on, too-small face masks, and orders that were never delivered. Pinduoduo has responded by taking down millions of listings representing fake-goods sellers on its app, and it has promised to cooperate with regulators and vet the app's sellers and products.

Though dismissing Pinduoduo as a lightweight, Alibaba is taking it seriously by recently launching Taobao Tejia or Special Deals, an app targeting its new rival's customer base with discounts for China's more price sensitive users.

China Goes Mad for Online Shopping

China is undergoing an e-commerce revolution, a symbol of capitalist success in the Communist country. The market is huge and fast growing. China has become the largest e-commerce market in the world, $1.1 trillion in 2018, and will reach a projected $1.8 trillion by 2022, more than double the US market of $713 billion. There's still plenty of potential. Only 38 percent of China's population shops online.[5]

The growth opportunity has spurred technology and e-retailing innovations and market battles. Dominant Alibaba and JD.com are fighting to retain their positions while newcomers grab a specialty area of e-commerce. Alibaba has debuted restaurants with robot servers, automatic vending machines, cashier-less stores, a secondhand marketplace with verified goods, and a channel for pampered pet products. Breaking other boundaries, Alibaba recently digitized a ladies' room in a Hangzhou shopping mall where women, waiting their turn to use the lavatory, can sample cosmetics and pay for makeup with a click on their smartphones, powered by Alibaba's Alipay.

E-commerce giant JD.com, which buys inventory from brands and operates its own logistics chain, is also exploring technology boundaries. What's new at JD.com is a digitalized supermarket I visited, 7Fresh, where shoppers pay with facial recognition, compare prices with instantly updated electronic tags, and check shipments of fresh foods to the store every step through blockchain tracking. Among other JD.com innovations are unmanned convenience stores, augmented reality mirrors for online shoppers to try on lipsticks and other cosmetics virtually, and a service where a well-dressed courier wearing white gloves drops off purchases at appointed times to car trunks or other designated spots.

Digging deeper into e-commerce, JD.com has embarked on a strategy to provide technology and logistics smarts as a retail service for its merchants and other retailers. This new initiative includes robotics-operated fulfillment centers, big-data analysis of merchandise sales, and supply-chain management. The company's logistics operation, which I saw at work in a Beijing warehouse, is pioneering with full automation. The operation spans to drones to reach rural areas, autonomous delivery vehicles on Beijing university campuses, and self-driving trucks on select routes.

Tencent has likewise begun digging into e-commerce too in China. Tencent bought into Alibaba's main rivals, purchasing 20 percent stakes in e-commerce giant JD.com and two hot startups that went public in 2018: Meituan and Pinduoduo. Tencent also gained control of consumer electronics e-retailer 51buy.com.

Lots of other power-packed moves are going on. Tencent and JD.com teamed up to co-invest $863 million in late 2017 in China's flash-sales site VIPShop, a seller of discount clothing. Both Tencent and Alibaba financed Chinese popular cross-border social commerce app Xiaohongshu, also known as Little Red Book but seemingly no connection to Chairman Mao's quotations book. Little Red

Book sells cosmetics and fashions to 200 million customers and has a following for its product and shopping experience review notes written by regular consumers and key opinion leaders. Both Tencent and Alibaba are making the most of their new reddish shopping app: when Tencent led its $100 million investment in 2016, Xiaohongshu promptly introduced a minishop on WeChat. In turn, when Alibaba led a $300 million financing two years later, the company integrated Little Red Book reviews into its Taobao shopping marketplace. Not left out of the e-commerce race, Chinese search leader Baidu began funneling shoppers hunting for items to JD.com.

Leading US companies are jumping into the Chinese e-shopping market too. Giant retailer Walmart bought into JD.com and in 2016 upped its stake to about 11 percent as its second-largest shareholder. The US chain has enlisted JD.com to fill Walmart customer orders in China. Google invested $550 million in 2018 for a tiny stake in JD.com as part of a strategic partnership to promote JD.com commerce on Google's shopping service.

It's unlikely that Chinese and American e-commerce giants will duel much. They're operating in separate spheres.

Amazon's efforts to penetrate the Chinese market have largely failed. Amazon entered China in 2004 by acquiring China's largest online bookseller, Joyo. Bezos changed the name to Amazon China in 2011, but management made classic mistakes: not letting local management make their own decisions, not realizing how price sensitive Chinese customers are, and failing to offer mobile payment options for China. Amazon has actually become Alibaba's customer, opening a store a few years ago on Alibaba's Tmall platform, popular for Western brands reaching Chinese customers, and paying Alibaba to run it.

Meanwhile, Alibaba and JD.com, while offering cross-border e-commerce platforms for international marketers to tap the Chinese market, haven't penetrated Amazon's stronghold in the United States. A few years back, Alibaba launched a boutique online shopping site, 11Main.com, designed to take a crack at Amazon with a niche store experience. Alibaba underestimated logistics support and concerns about product selection, quality, and safety. Its online marketplace model, which connects buyers and sellers rather than selling directly to consumers as Amazon does, wasn't familiar to Americans, and sales were lackluster. One year after its start in 2014, Alibaba sold the failing 11Main.com to US social shopping site OpenSky, which it now owns.

Alibaba has shifted its priority in the United States to working with American businesses to sell to Chinese consumers. Leader Jack Ma traveled to Detroit in June 2017 and staged a rally, a variety show, and a training session to convince US small business owners to try its online marketplace platform to sell goods—and fulfill his promise to newly elected President Trump to create new jobs in America. The event in Motor City's Cobo Center attracted more than 3,000 attendees—a door opener for Alibaba in the United States. About 7,000 US businesses, mostly large companies, sell products to China on Alibaba.

Alibaba vs. Amazon

Alibaba today is sometimes compared with e-commerce force Amazon although their business models and orbits differ—though both founders have ironically become newspaper owners. Jack Ma paid $266 million to buy Hong Kong's leading English-language newspaper the *South China Morning Post* in 2015 while Jeff Bezos bought the *Washington Post* for $250 million in 2013.

A comparison of the China and American leaders of e-commerce, Alibaba and Amazon, reveals how they've followed parallel paths into diversified market segments and into a wide array of online shopping and related services, but from different bases, time lines, and strengths. The biggest difference between Amazon and Alibaba is the underlying business model. Amazon is an online retailer while Alibaba is a digital platform connecting customers with sellers directly and owns no inventory or warehouses.

Alibaba started in 1999 five years after Amazon as a business-to-business e-commerce site in China. Its earliest US rival was not Amazon but eBay, which Alibaba crushed with its consumer auction site Taobao in 2006 after a vicious battle. Amazon bought its way into China and rebranded as Amazon.cn, but it's nowhere near its giant-sized presence in the United States.

Both of these leading e-commerce companies have moved into financial services, cloud computing, enterprise chat, payments, and online video—which all feed into making e-commerce more effective and efficient. Alibaba got into health care, financial services, gaming, enterprise chat, and physical stores a few years ahead of Amazon. Meanwhile, Amazon beat Alibaba to e-commerce cloud services and online video by four to five years and was way ahead in hardware with the Kindle in 2007. Alibaba has moved into online travel, while Amazon hasn't. "The delivery speed and range and quality of new services at Alibaba is as good as, or better than, Amazon," says Hans Tung of GGV Capital, the venture firm that backed Alibaba in 2003 when very few venture firms on Sand Hill Road would fund a Chinese company.

Young, Rich, Serial Entrepreneur

But this is old news compared to the story of the upstart Pinduoduo.

The social shopping app is the fourth digital startup in China for visionary founder Colin Huang, who launched his career in Silicon Valley at Google and returned home to try his hand at entrepreneurship. The son of factory workers, Huang, 39, is now among China's newest and richest billionaires, with a net worth of $12.4 billion.[6] Huang grew up in Hangzhou and earned degrees in computer science at Zhejiang University and the University of Wisconsin–Madison before starting at Google in 2004 as an engineer in Silicon Valley. In

2006, he relocated to China to help set up Google China, but—as the story goes—soon left after becoming frustrated with continually flying back to headquarters for approvals for seemingly minor changes, such as the color and font size of search results. In 2007, with a fortune made from Google stock, he struck out on his own. In quick order, he started online consumer electronics and home appliance seller Ouku.com, which was sold, then online goods marketing service Lequi and online gaming studio Xinyoudi in 2011. The idea for Pinduoduo came to Huang when he was recuperating from an ear infection, taking time off, and studying the success of Tencent and Alibaba. His venture investor friend and former Google colleague James Mi convinced him not to go for his original idea, which was a manicure booking service app. Instead, Huang smartly combined the gaming and social networking aspects of Tencent's business with the e-commerce focus of Alibaba and baked in some Groupon-type features such as team buying. In 2015, he raised approximately $18 million from Banyan Partners, IDG Capital, and Lightspeed China Partners and incubated Pinduoduo from his gaming company and another idea to sell fresh produce online. Soon Pinduoduo took shape as a full-fledged online marketplace selling all kinds of discounted items, including fresh fruits. Differing from earlier e-commerce businesses, Pinduoduo caught on among a large and fast-growing segment of Chinese consumers outside the major cities.

Pinduoduo relies on artificial intelligence and big data analysis at its core. Huang's team includes high-tech and data mining experts who have experience at Baidu, Google, Yahoo!, and Microsoft and who were his colleagues at his gaming startup. Showing his ambitions, he's recruited more star power: Tian Xu from Baidu as his vice president of finance and Lin Haifeng from general manager of Tencent's M&A department as a director.

Venture capital funding was tapped two times before Pinduoduo snared a $3 billion investment in 2018 from Tencent at a valuation of $15 billion. That large infusion led to its IPO on Nasdaq in July 2018. Six months later, Pinduoduo raised $1.4 billion in a second-ary offering to fuel growth. Revenues increased fivefold in 2018 to $1.9 billion and are projected to double in 2019 to $4.15 billion, but fast-paced expansion is costly, with operating losses of $1.5 billion in 2018.[7]

On his board of directors are high-powered venture capitalists Neil Shen of Sequoia Capital China and Zhang Zhen from Gaorong Capital and previously IDG Capital. Huang and his founding team retain a controlling 50.1 percent share of the company.

Table 8-1

At a Glance: Pinduoduo

Founder: Colin Huang

Launched: 2015

Location: Shanghai

Business: mobile shopping app

Status: Nasdaq IPO raised $1.7 billion at $24 billion valuation in July 2018

Notable: pioneered a mobile shopping model with social teams earning group discounts

Financing: Tencent owns an 18.5 percent stake

Source: company reports

Pinduoduo founder Huang feels the pressure of running the dynamic startup that's turning e-commerce in a new direction. He didn't go to the Nasdaq bell-ringing in the summer of 2018 himself but sent a customer representative who had been lucky in one of the prize drawings, and he held a separate IPO ceremony in Shanghai. In his message to IPO investors, he wrote, "Pinduoduo's survival depends on the value it creates for its users; I hope our team wakes up feeling anxious every day, never because of share price volatilities but because of their constant fear of users departing if we are unable to anticipate and meet users changing needs."[8]

CHAPTER 9

THE DETROIT OF ELECTRIC VEHICLES: CHINA

Buy an electric car in China, and you'll receive a free license, a $10,000 subsidy, and access to charging stations. China is the world's leading market for all-electric vehicles, led by Tesla challenger NIO and Alibaba-backed Xpeng Motors.

China is on a journey to disrupt the automobile industry with electric cars, self-driving tech, and seamless internet connections in a race with the United States to own the car of the future. Stand guard, Detroit and Silicon Valley.

Chinese electric car innovator Xpeng Motors is at the forefront of a massive upheaval of the 100-year-old auto business. I interviewed the founder of this highly charged company at its recently opened research base in Mountain View, not far from the Computer History Museum, in a standard, low-rise office building that shows no trace of China. Xpeng Motors was about to showcase a new all-electric model for the first time stateside at an event held by its US partner for autonomous driving, Nvidia, a supercomputing artificial intelligence leader. At about the same time, Tesla and Apple were charging two Xpeng engineers of stealing their proprietary technologies.

Xpeng Motors is named after the founder He Xiaopeng, who made a fortune in selling his first startup to Alibaba and now is deep into ramping up his auto tech startup. The Guangzhou-based company is headhunting experts in autonomous driving and machine learning, opening its own factory in the metropolis of Zhengzhou, and marketing its G3 sport utility vehicle priced in the midrange of $35,000 to China's young, tech-savvy drivers. The SUV is packed with fun, connected technologies such as in-car karaoke and a 360-degree rotatable roof camera for taking group selfies. As many as 10,000 customers in China ordered the sporty vehicle within three months. Xpeng rolled out a coupe, code-named E28 with higher performance and more advanced self-driving features, at the 2019 Shanghai auto show, and is on track to sell at least 40,000 electric vehicles in China in 2019—more than Tesla's Chinese sales. Xpeng has beefed up sales with an online-to-offline method to increase awareness and generate customer leads through Chinese social media WeChat and Weibo, and it has invited prospects to take a peek at pop-up stores in shopping malls or at its nine direct outlets in key locations, soon expanding to 70 in 30 cities.

This is China speed and technology disruption wrapped in one. Over the past 20 years, I've seen China streets evolve from bikes, carts, and public buses to VWs, Buicks, Audis, and, next, electric vehicles. Within just the past few years, China has gotten ahead of the curve in adapting to new energy from gas-powered vehicles. China has emerged as the largest and fastest-growing electric vehicle market in the world. Baidu, Alibaba, and Tencent are all major players in this ring, funding select startups such as Xpeng and NIO and using their technology smarts and power to build fun, connected, and intelligent vehicles of tomorrow.

China Electrifies the Car Market

At Chinese startup Xpeng Motors in Mountain View, where the high-end R&D for autonomous driving takes place, founder He told me, with the help of a translator, about his longtime dream to create something "innovative and impactful, that will bring big changes to the world." One idea he considered was building a town in the ocean, but he realized technical issues such as power supply and garbage disposal would be too difficult.

Developing an intelligent vehicle with autonomous driving features doesn't seem much easier to do than creating an oceanic town. The company's chairman and CEO outlined his accomplishments since embarking on this journey in mid-year 2014, just a few months after he sold his first startup and began investing money into his new dream for the road. Born in Hubei province to parents who both worked as technicians, the uber-connected He, 41, strikes me as pretty calm considering the enormous challenges ahead, far more complex than with his prior internet startup.

Having enough supercharging stations to keep its vehicles juiced up is just one hurdle. Xpeng intends to have 1,000 Xpeng supercharging stations nationwide by 2022 and partner with third parties to bring on 100,000 charging spots. Xpeng is providing free installation service for at-home charging too. Its coolest innovation is a quick battery-swapping technology system. Drivers stay in the car while the battery is exchanged within three minutes. Xpeng will not make its own battery cells—those are sourced, but the battery pack is designed from in-house technology.

The founder of Xpeng has called on Alibaba and other powerhouse investors to come along for the ride. Prominent strategic investors are Alibaba, contract electronics maker Foxconn, and Chinese ride-hailing startup UCAR. Leading venture and private equity investors are Morningside Venture Capital, IDG Capital, Hillhouse Capital, Primavera Capital Group, Matrix Partners China, and GGV Capital.

His mentor, famed angel investor and Xiaomi founder Lei Jun, is a backer too. So are several other internet tycoons from China, including Jack Ma–backed Yunfeng Capital, former high-ranking Tencent executive Wu Xiaoguang, and Li (David) Xueling, the founder of video-based social network YY. At Xpeng Motors, it's clear that China's tech elite come close to matching the clubby investor and entrepreneur circles of Silicon Valley.

As much as $1.5 billion has been invested, and more capital is being raised. An IPO could be in the offing, but no timetable has been set yet. That will be a job for former JPMorgan Asia investment banking chief Brian Gu, who has been hired at Xpeng as vice chairman and president.

China has shifted into high gear in making and selling new energy models after largely missing out on making much of a dent with gas-guzzlers. As China pulls forward, the United States is not in the driver's seat. Motor City stands to lose more ground as the auto-making capital of the world—the US Big Three all have their own electric vehicle ventures in China. Silicon Valley's continued competitive edge in auto tech is looking uncertain. China is producing more than half of the world's electric car batteries but hasn't yet caught up to the gold standard Tesla with a 300-mile driving range. Tesla plans to rev up limited sales in China in a few years by churning out cars from its own factory in Shanghai and bypassing high tariffs.

Competition is getting keener between the United States and China, which together account for half the electric vehicles coming off the assembly line globally. Pressures have increased as the US-China trade war continues as a major geopolitical issue and a US crackdown on Chinese corporate espionage intensifies. Recent lawsuits from Tesla and Apple claiming theft of their trade secrets by engineers working at Chinese e-vehicle (EV) makers could be the tip of the proverbial iceberg.

China has a lot to overcome before emerging as champion of

next-generation cars. Chinese-made Geely, BYD Auto, and Great Wall Motors cars aren't universally known, and Chinese cars suffer from an image of cheap and crappy. Fifty years ago, Japanese automakers were ridiculed for producing small cars of questionable quality, but look at all of the Toyotas, Hondas, and Nissans on the road today. This could be the time for China's new electrics to leap forward and dominate the smart-car market of the twentieth century, but there are many ifs that could stand in the way.

The development of China's EV industry is an element of the nation's Made in China 2025 policy to own key industry sectors. But the Chinese government recently scaled back subsidies of electric vehicle sales by half in an attempt to spur makers to rely on innovation rather than government assistance, and it plans to completely end subsidies by 2020. This subsidy reduction will thin the ranks of some 50 electric vehicle makers that have emerged in China since 2014. Moreover, the US–China tech war could stymie global expansion of Chinese EV makers.

Techies Lead China's Car of the Future

Two Chinese car startups are at the forefront of China's new and fast-growing EV market: Xpeng Motors from the southern Chinese city of Guangzhou, and a Tesla wannabe, NIO, from Shanghai, which in September 2018 became the first Chinese e-vehicle maker to go public on the New York Stock Exchange.

Their founders are not car buffs but techie serial entrepreneurs, and wealthy ones at that. The founder of Xpeng sold his prior startup UCWeb, a leading mobile web browser, to Alibaba in a $4 billion cash and options deal in 2014 that was the biggest acquisition in Chinese internet history. The developer of NIO, William Li, took his previous startup, web content provider Bitauto Holdings, public on the NYSE in a 2010 IPO that raised $127 million. See table 9-1.

Table 9-1

At a Glance: NIO

Founder: serial entrepreneur William Li

Location: Shanghai

2018 Financials: Revenues $720 million; **Losses** $1.4 billion

Milestone: took NIO public on the NYSE and raised $1.2 billion

Notable: electric vehicle is known in China as a Tesla killer

---

These Chinese car startups are relying on their tech smarts, capital connections, government support, and global outlook as a blueprint for tomorrow. The DNA of their companies is tech, not auto making. Since the early 1980s, China has wanted to have its own powerful automobile industry, but it's not been until now that the nation can realize that goal through its leading technology companies, notes Michael Dunne, CEO of Hong Kong–based auto tech advisory firm ZoZo Go.

Now these homegrown tech-oriented startups are powering up with electric and autonomous vehicles that can make China proud. With no legacy of gas-powered engines or Chinese state-owned, auto-making enterprises, they have an open road to roam in China.

"China has the potential to make world-class vehicles with their smart and cashed-up tech companies," said consultant Dunne, speaking at an Asia Society meeting in Northern California. "By relying on tech companies instead of the entrenched automakers, they have found their point of leverage. They can make a great leap forward with scale, regulations, technology, and a capital river of billions. I think the US is in trouble."[1]

So far, no Chinese electric vehicle brand has made a splash in

Western markets and established trust with customers as the Japanese and Korean makers have. Rising protectionism and growing distrust of Chinese technology in the United States don't make it any easier for Chinese makers to break in.

"Across the various spectrum of industry from telco, entertainment, and even real estate, China has learned that entering the US market, a free market driven by consumer demands, is challenging," said Eric Mika, who leads government affairs and business development for Los Angeles–based Canoo, a creator of subscription-only electric vehicles secured by blockchain technology that is heading toward a US launch in 2021 and China shortly afterward. "Add in the business cultural differences and the perception of 'Made in China' versus 'Made in the USA,' I assume that the Chinese EV industry entry into the US market and Euro Zones will be challenging."

There's an enormous opportunity for China's new-energy, pollution-fighting cars right within their own homeland—at least for the strong that can weather China's economic slowdown, competition, and the end of subsidies for electric car buyers. The two startup leaders, China's Xpeng and NIO, are both focusing first on China before any rollout to Western markets, although both have aims. NIO used its New York IPO for positioning itself as a global competitor and a challenger to Tesla.

Xpeng plans to expand to regional markets Hong Kong and Singapore by 2020 and perhaps Western Europe and the United States later on. "We want to do well in regional markets first. Expanding to global markets too early will dramatically increase costs," he said. "We'll start in the Chinese market, establish our brand and presence there, and build out sales and support services before we think about the US market a few years from now. Expanding to global markets too early will dramatically increase the cost base."

Electric vehicle sales in China are surging, by 62 percent in 2018 to 1.3 million, about 4 percent of the total Chinese car-buying market compared to 2 percent of auto purchases in the United States.

American automakers are investing billions in electric vehicle technology, and China is their biggest potential market. The US market is only about two-thirds the size of China's EV sales. Americans like the idea of all-electric cars but are turned off by impracticalities. Driving ranges are typically limited to around 200 miles for pure electric cars, and there are too few supercharging stations. The Nissan Leaf and Chevrolet Bolt are most typically driven in neighborhoods for errands, work commutes, and as a second car.

Nevertheless, the electric future is here and getting bigger and better. By 2022, 10.3 million new electric vehicles are projected to be sold globally, with China expanding by 37 percent to 3.6 million and the US increasing 26 percent to nearly 2 million.[2]

Funding is plentiful for China's brand-new electric vehicle producers. Over the past three years, venture capitalists have poured in $14 billion while China's tech titans have come on board: Alibaba is a major investor in Xpeng while Tencent and Baidu are investors in NIO.

China's Futuristic Car

The car of tomorrow is not just about getting from one place to another. It's a comfortable personal space with lifestyle and social features—and a new source of business revenue and profits for automakers. This vision is becoming a reality as truly self-driving cars, not just self-parking models, arrive by 2025, freeing drivers from the steering wheel. "Our

> *"Our long-term goal is to become a global intelligent mobility company."*
>
> **He Xiaopeng**
> Founder, chairman, and CEO,
> Xpeng Motors

long-term goal is to become a global intelligent mobility company. Tomorrow, the car will become an entrance to a mobile ecosystem around a platform of multiple services to our customers," said Xpeng founder He.

China's electric car startups have built-in artificial intelligence systems that follow voice commands and recognize faces to activate in-car entertainment systems of music, games, and karaoke as well as customize settings for air conditioning, seating, and radio volume. Their AI interfaces have names: Xmart at Xpeng and Nomi at rival NIO. Smart driving features include auto parking but not true self-driving yet, real-time navigation, mapping of charging facilities, and sensors for speed limits, emergency braking, lane changing, and warnings of possible collisions.

The Car Chase by Google, Baidu, Ford

China is projected to be the market leader of intelligent, connected self-driving vehicles, projected to reach $14 billion by 2020. Google's self-driving cars are tested on Highway 101 and have logged more than 5 million road miles in preparation for launch in 2020. Baidu has teamed up with Ford and Volvo to test and accelerate China launches of self-driving vehicles. China is getting ahead in mass adoption of autonomous driving with robo-taxis, city buses, and driverless private cars taking over most of China's vehicle market by 2027, leading up to a $1.1 trillion market from mobility services and $0.9 trillion in sales of autonomous vehicles by 2040.[3]

China's leading electric car startups are laser focused on building a research foundation for self-driving. China-anchored Xpeng is expanding its R&D team in California from fewer than 100 to 150 by the start of 2020 to work on proprietary breakthroughs, hiring from Silicon Valley tech leaders as well as universities, Xpeng's founder told me when we met in Mountain View.

Besides the California R&D dream team, Xpeng has a sizable research group in China working on AI, autonomous driving, and connected vehicle technologies which is being expanded to 800 this year. Altogether by the end of 2019, Xpeng is counting on 1,000 people in R&D globally for these advanced technologies and another

2,000 on power train and auto hardware testing. Nearly two-thirds of the Xpeng current staff are in R&D, split among five research centers. As Xpeng staffs up other positions, the goal is to maintain R&D levels at 50 percent of the overall head count. "Having R&D centers in the US is not only important for understanding the US market but also for understanding the broader Western markets such as Europe," CEO He told me, noting for example that safety requirements and regulations for road safety are very different in China than in the United States.

In California, Xpeng is aggressively head-hunting top engineers to build out its research and development team working on autonomous driving and artificial intelligence technologies. Our interview in early March 2019 was a few weeks before Tesla charged a former engineer recruited by Xpeng of stealing its trade secrets for driver-less technology. Xpeng is not accused in the suit and has responded that it wasn't aware of any wrongdoing but has launched an internal investigation. The lawsuit is a clear signal of the US-China technology race and edginess by Silicon Valley tech leaders that are seeking to protect their crown jewels and defend their turf.

High-profile recruits are finding their way to Xpeng. There's the new vice president of the autonomous driving unit, Wu Xinzhou, who previously led the self-driving team at Qualcomm. Then there's Gu Junli, VP of autonomous driving, the former tech lead of Tesla's machine learning road map for its autopilot system. The Xpeng engineer Tesla has accused of stealing files and codes, Cao Guangzhi, worked on Tesla's autopilot team. Xpeng's director of production quality, Miyashita Yoshitsugu, is from Toyota and is a leading expert in the field of lean production and zero defects.

Silicon Valley tech companies are out to prevent their precious intellectual property from being stolen, a troubling issue as talent moves from one company to another, notably among US and China rivals. Tesla's lawsuit seeks return of its proprietary information, knowledge about how the company is using it, and damages from alleged theft. In a WeChat post, Xpeng founder He called the lawsuit "questionable" and noted that both his startup and Tesla are innovators and that a "flow of talent" between companies is normal. Xpeng separately issued a press release stating the company's respect for intellectual property rights and confidential information, noting it was not aware of any alleged misconduct by the engineer but is investigating internally.

In an earlier incident linked to Xpeng, a hardware engineer is facing criminal charges by the US Justice Department after Apple accused him of downloading files containing proprietary information in advance of joining Xpeng. He was arrested at San Jose Airport on his way to China, and Xpeng subsequently fired him. In another case involving a China autonomous vehicle company, Apple charged an engineer also leaving for China of stealing its driver-less car secrets.[4]

NIO in Road Race with Tesla

From another corner of China's fledgling electric market, premium maker NIO is coming on strong as a Tesla challenger with *Star Trek*-y looking cars and a starting price tag of around $70,000, much less than Tesla in China. The founder, William Li, has been called the Elon Musk of China. His company's slogan is "Blue sky coming."

Tech-savvy Chinese customers can link their mobile phone to NIO and tap on a screen for repairs, maintenance, and quick power boosts by battery swapping and mobile charging vans. Like at Xpeng, Li has poured some of his own money into the startup, which is backed by China tech titans Baidu and Tencent plus veteran investors Sequoia Capital, Hillhouse Capital, and Temasek. One of NIO's selling points is a clubhouse for users in prime real estate locations in major Chinese cities. Its first SUV, the ES8, had sales of approximately 10,000 cars. A second, the ES6, priced at $52,000 is planned later in 2019.

In going public in New York in September 2018, NIO positioned itself as a global EV maker and planned to sell cars in the United States. But no mention of a US launch has been made yet and there are signs that NIO is lowering its sights. The company, which had a $1.4 billion loss and revenues of $720 million in 2018, recently scrapped plans to build its own factory in Shanghai. NIO will continue to contract out production to a Chinese state-owned plant.

Several other Chinese automakers have a running start in China and aim to crack the US market. One is Hong Kong–based Byton, which plans to start selling its cars in China soon, followed by rollouts in the United States and Europe by late 2020. Its electric sport utility model, M-Byte, has drawn favorable comparison with Tesla models.

Another, Chinese-owned Karma Automotive in Southern California, is recharging its Revero gas-electric sedan. Priced at $130,000, the Karma Revero has proven popular with the Hollywood set for its flashy design and eco-friendly features, but it's been criticized for being impractical for short driving ranges of only 50 miles, supplemented by a four-cylinder gas motor.

The Entranze electric car from China, made by Guangzhou Automobile Group and designed in Los Angeles, plans to start selling in the United States in 2020.

The Chinese keep coming with ambitions to build up their global standing. Auto consultant Dunne estimates that there are as many as 175 Chinese auto and auto tech locations in the United States.

The once promising Faraday Future, however, is one that hit a speed bump. Launched by Chinese tech and digital media innovator LeEco with grand plans for a Nevada plant and seamlessly connected e-vehicles, Faraday faced a cash crunch. Operations were moved to Los Angeles thanks to a lifeline from a new Chinese investor, but Faraday has been struggling with more financial setbacks and a string of lawsuits from unpaid suppliers.

Geely and Great Wall Left in the Dust

China's grand ambition to own the electric car market is a major advance from 20 years ago, when the nation began pursuing a goal of making its own cars. The cornerstone of its industry then was government-mandated, fifty-fifty joint ventures between state-owned enterprises and foreign automakers to get car-making knowhow, such as GM's partnership with Shanghai Automotive Industry Corp. to sell Buicks, Chevrolets, and Cadillacs in China and Ford's joining with Changan Automobile to produce and sell Escorts to the Chinese. Three privately held Chinese auto companies ramped up in the 1990s: fast-growing Geely, which bought Volvo from Ford in 2010; electric vehicle maker BYD, which is funded by Warren Buffett and Bill Gates; and Great Wall Motors, which is China's largest SUV and pickup truck maker. By 2000, China was selling 1 million cars annually. Ten years later, China became the world's largest car market with 10 million sales in 2010. But China still did not have a world-leading car brand, air pollution was worsening, and China was becoming dependent on Middle Eastern oil.

Enter China's electric car makers beginning in 2014—and a realization by Beijing that the only way to take leadership of an entirely new segment of the auto industry "is to subsidize the heck out of it," said China auto tech consultant Dunne.

To accelerate the expansion of e-vehicles, the Chinese government has doled out subsidies of $10,000 for new buyers and handed over rebates for obtaining driver licenses that can cost upward of $12,000. The government has also pledged to build 12,000 centralized charging stations, including 800 intra-city fast-charging stations by 2020. Its new $47 billion fund for high-tech industries pinpoints electric vehicle development. Cutting emissions and cleaning up air pollution are clearly on China's policy agenda, as the country bids to become a world power in electric vehicle technology as outlined in President Xi Jinping's "Made in China 2025" initiative.

The Chinese government is requiring all global automakers in the country to make 10 percent of their vehicles electric models by 2022. There's even a possibility that gasoline-powered cars could be banned in China. Public buses in Shanghai are all electric, and a driver-less bus has been put in motion in a park in Wuhan.

China's tech titans have joined this transportation revolution. Tencent has partnered with Changan Automobile Company to develop an "internet of vehicles." Baidu is producing autonomous minivans in partnership with Chinese bus manufacturer King Long.

Producing new electric models in China is a path forward for GM, Ford, and Chrysler as well as Nissan, VW, and Toyota. In 2018, passenger vehicle sales overall in China were down 4 percent—the first decline since 1990—in an uncertain economy and a crackdown on lending.[5] Sales of Ford and Chrysler in China plummeted in 2018 while General Motors volume did well with Cadillac.

Ford has entered into a new joint venture to produce electric vehicles in China and plans to start producing self-driving vehicles in Michigan by 2023, investing $11.1 billion overall in electrified models. GM is ending production of its plug-in mass-market Chevrolet Volt and moving upscale with Cadillacs, seeking to catch up with Tesla and leading its push into an electric future. Fiat Chrysler is adding a plug-in Jeep Wrangler and an SUV to its lineup in 2020. Honda has a new division dedicated to making fully electric cars, has been experimenting with hydrogen-fuel-cell vehicles, and is partnering with Chinese companies. Toyota's first electric-only cars will go to China first, then to the United States and Europe.

Enter Tesla to China

And what of Tesla? I remember when Tesla luxury sedans arrived to the Chinese market and my group, Silicon Dragon, showcased the new model outside our event venue in Shanghai. A crowd gathered and marveled at its beautiful design and features. Inside, a Tesla executive told us of plans to sell cars to China's newly affluent, status conscious consumers. But Tesla got a slow start and has sold only an estimated 30,000 Tesla models since launching in China in 2013, compared to about 180,000 in the United States, where it is the dominant seller with about half of EV sales.

The issues for Tesla in China have mainly been high prices due to tariffs and too few supercharging stations. Tesla's China sales slid in 2018 in the wake of the country's slowing economy, tariff swings, and multiple price shifts in the US-China trade war. To the rescue, visionary founder Elon Musk. He recently flew to Shanghai and shook hands with the mayor in a ceremony to debut Tesla plans to construct its own megafactory in the city, which could crank out 500,000 cars annually, as much as its Fremont, California, plant. With local production in China starting in a few years, Tesla could get around 25 percent tariffs on its models, which are priced at $78,000 to $150,000 for high-end sedans.

As China's government ends subsidies for Chinese electric carmakers and Tesla bypasses high tariffs with China production, the electric vehicle market could become a much more level playing field for US and Chinese rivals.

CHAPTER 10

THE AGE OF DRONES AND ROBOTS

China has turned big-time to drones and robots for handling lots of tasks humans can't or don't want to do. Chinese drone startup DJI is the world leader, EHang has a passenger-carrying drone, and robotic vacuum cleaners and window washer startups are getting accelerated at HAX in Shenzhen.

Drones made headlines in the United States when Amazon CEO Jeff Bezos said his company would deliver packages to a customer's doorstep within a half hour. That was five years ago, and these drones have not taken flight, held back by federal regulations restricting package delivery except for a few pilot programs. But in China, retailer JD.com and delivery service Ele.me are already testing delivery drones to cut costs and reach rural shoppers.

China's pioneering drone maker DJI Innovations is getting attention today—not always in a favorable light. DJI made news when an inexperienced pilot famously crashed one of its drones on the White House lawn. And it made more news when the US Army restricted soldiers from using drones made by China's DJI, based on claims that the Chinese drone maker was collecting data from its unmanned flying machines in the United States and relaying that information back to China. DJI denied those claims and alleviated some issues by issuing a software update that prevents its drones from flying within a 15.5-mile radius of downtown Washington and introducing a privacy mode on its drones to prevent transmissions. But the US military ban remains.

It's not just a security issue but an economic one. Beijing aspires to become the world leader in several technologies, including robots and drones, as part of the country's "Made in China 2025" strategy to reduce its dependence on imports, develop elite talent, and improve the country's production systems. China has made swift progress in developing a thriving robotics and drones industry as an upgrade of its manufacturing and military sectors, already surpassing Japan as the world's largest market for industrial robotics, and is on the way to more than one-third of commercial robots in use worldwide. Chinese state-owned conglomerates, emerging companies such as DJI, and venture capital firms are acquiring and investing in robotics technologies abroad. Such growth and ambition could jeopardize many US technological advantages, a research report prepared for the US-China Economic and Security Review Commission concludes. The report recommends the US government promote advanced manufacturing and robotics technologies, monitor China's advances, review bilateral investments and cooperation, and consider closer vetting of academic and research exchanges.[1]

Globally, the $115.7 billion robotics and drone market is projected to nearly double to $210.3 billion by 2022.[2] Of the $103.4 billion robotics market worldwide, China will be the largest market, at about one-third of spending, followed by Asia, the United States, and Japan. For the $12.3 billion drone market, the United States will be the biggest market at about 40 percent, trailed by Western Europe and China. The takeaway here for Sinophiles is that the Chinese market for both robotics and drones is expected to grow the fastest in the world over the next few years to 2022, increasing by 24.6 percent and 63.5 percent consecutively.

Growing confidence in the usefulness of robots and drones

has led to an influx of venture funding over the past few years. These two game-changers are starting to replace humans in basic chores, industrial-size jobs, emergency responses and disaster relief, health-care deliveries to rural areas, weather forecasting, waste collection, construction planning, aerial real estate photography, and eldercare.

Chinese startups in robotics and drones are coming by the dozens. China-made UBTech in Shenzhen recently pulled in $800 million in funding led by Tencent for its AI-enabled, human-looking smart robots that can guide guests through hallways, walk on uneven terrain, respond to voice commands, and teach kids to code and build robots, becoming an instant unicorn, valued at $5 billion.

Another to keep in your sight lines is Horizon Robotics, a developer of artificial intelligence chips for robots and self-driving vehicles, which recently snagged a $600 million investment led by South Korean conglomerate SK Group on top of funding from Intel Capital and a who's who of China venture capitalists.

HAX Accelerates in Shenzhen

On the same day that I was stopping by DJI for a look-around and interview at their Shenzhen location, I managed to squeeze in a tour of the HAX accelerator, a US-funded investor and venture builder of hardware and robotics startups for consumer, enterprise, and industrial markets.

In its batch of startups, HAX has funded and accelerated more than two dozen robotics startups. A rising star is MakeBlock, a maker of do-it-yourself robotics kits and facial recognition technology for robotics engineering that has attracted funding of $77 million from Sequoia Capital China and other majors and could go public soon. Others backed by HAX are made for unglamorous industrial work: Elephant Robotics makes a flexible robotic arm for small business assembly lines, Rational Robotics does autonomous painting, Avidbots cleans floors commercially, Plecobot washes skyscraper windows, and Youibot inspects vehicles for safety.

I observed as several founders practiced their investment pitches on a demo day at HAX in Shenzhen. They were gearing up to raise capital after HAX invested seed money of $250,000 for 15 percent equity in exchange for providing startup space and expertise on start-to-finish projects from sourcing and supply chain to prototyping, design, engineering, and manufacturing. HAX is right next door to Hardware Alley, the ultimate supply chain with representatives at small booths for many Chinese electronics component makers. The HAX labs, workshop, and mentoring outfit was set up in 2012 by visionary entrepreneur and investor Sean O'Sullivan, who lives in Princeton, New Jersey, and is the inventor of Nasdaq-listed MapInfo, which pioneered the mapping software that Google uses. Putting his money to work in fueling tech startups, HAX accelerates about 50 hardware companies per year from Shenzhen and San Francisco locations and is part of the SOSV venture capital umbrella, which spans to startup accelerator Chinaccelerator in Shanghai and MOX in Taipei for mobile-only startups run by startup evangelist and investor William Bao Bean. Its biggest hits so far are electric bike-sharing startup Jump Bikes, acquired by Uber for $200 million, and Bluetooth earphone maker Revols, sold to Logitech.

Robots at Work

Many robots today are finding all sorts of chores to do, like Pepper, the humanoid robot made in France that greets retail store visitors, and Roomba, the US maker of self-operating vacuum cleaners. Others end up in warehouses, carting boxes. Then there's the touch-sensitive dog robot Aibo made by Sony.

These rather tame robots are hardly the types identified by the US federal agencies as security threats, but since many are invented, funded, and accelerated in China, they do represent an economic challenge to US leadership of a major tech sector.

Chinese Drone Maker Flies High

Next, I found my way to drone maker DJI, about a 20-minute taxi ride from HAX and not far from Tencent's new headquarters. Flying high, DJI has soared to take two-thirds of the drone market globally. Still privately held, DJI has rapidly grown into a drone master with a global workforce and 11,000 employees. No drone company in the West comes close to DJI dominance. It's the rare Chinese company that can ace a market globally, and DJI has done that by leapfrogging others in China and the West with its fine-tuned innovations, technical superiority, speed, and efficiencies.

I had a memorable visit to DJI in the southern Chinese city of Shenzhen at the sprawling Viseen Software Tech Park and got a demo of its heavy-duty and lightweight drones flying high above the surrounding corporate buildings. Those drones whirl by like giant bumblebees but are actually hardworking aerial robots that can do surveillance and inspections for utilities, construction sites, airplanes, and trains from onboard cameras. Also for hobbies and fun, drones are popular for capturing perfect images of weddings or just playing in the backyard.

David Benowitz, a manager in the company's enterprise department, showed me around DJI's sleek showroom and detailed the types of drones that DJI makes for industrial and consumer markets, including its well-known Phantom recreational drone, which retails for about $1,000.

Amazingly, we discovered that we both grew up in central Ohio; our high school sports teams were archrivals. Benowitz, who describes himself as a sci-fi geek, came to Shenzhen to intern at DJI two years ago, not looking back much at his hometown or Georgetown University, where he earned an economics degree. He's learned Chinese and fits in well with the company's youth culture—the average age of DJI's employees is 27.

Soon, DJI will be moving to a new headquarters that reflects the ambitions of its founder, Frank Wang, a press-shy product genius who was inspired by Steve Jobs' dictum: design the product first and see how the market responds. DJI's new flashy home in Shenzhen is a futuristic twin skyscraper designed by Foster & Partners, the same architect as for Apple's orbit-like base in Cupertino. The plush building features cantilevered floors, a sky bridge where drones will be tested, and even a robot-fighting ring.

The Apple of Drones

DJI has positioned itself as the Apple of drones. Rumors have popped up that Apple would buy DJI as the iPhone maker mulls an entry into the drone market. In a stroke of marketing genius, DJI has obtained a dedicated section of prime real estate space at Apple retail stores globally for a line of advanced consumer drones. Its drones are sold on Amazon, eBay, Alibaba's online retail service AliExpress, DJI's site, and retail stores.

Taking another cue from Apple, DJI has moved into physical retailing with swanky stores. DJI has four of its own retail outlets, including a three-story flagship store with an Apple-white decor in Hong Kong. Here, customers can see DJI's line of professional and consumer drones, watch pilots fly drones, get technical assistance at workshops, and explore photos and videos captured by aerial enthusiasts.

Besides DJI, there really isn't any major player in the drone space, except one Chinese entrant that has been making a splash with a passenger-carrying drone. This is EHang, which wowed crowds at the 2016 Consumer Electronics Show with the first human-carrying drones—an autonomous aerial vehicle designed for personal transportation from point to point. EHang has since filed a bankruptcy in the

United States to close its US office and refocus on the China market and R&D. It's on a quest to commercialize the drone copter. Derrick Xiong, a cofounder of EHang, got funding for his initial innovation, the Ghost Drone, controlled by a mobile app, from a crowd-funding campaign that raised $600,000 on Indiegogo in 2014. GGV Capital followed with $10 million the next year. There are rumors that EHang may seek a public listing that could raise $500 million.

Model Plane Enthusiast Founder

It helps to thrive in this business if you love flying. DJI's founder Wang hails from Hangzhou and is the son of a teacher and small business owner. Wang had a dream of flying from when he was a kid, and he spent a good deal of his childhood building and flying model airplanes and wondering how to make a toy plane that wouldn't crash. While attending the Hong Kong University of Science and Technology as an engineering student in 2003, he got a research grant of $2,300 in 2005 to develop a drone. From his Hong Kong dorm room and with the help of his mentor, Professor Li Zexiang, he formed an unmanned minihelicopter flight control system that became the seed of DJI in 2006.

Originally a hobby and a student project, DJI turned into a full-fledged business. DJI's Phantom drone, introduced in early 2013, was the first minicopter that could be taken out of the box and assembled in one hour for flight, and without falling apart on its first crash. Soon, recreational drones became the latest fad and Wang's fortunes soared.

Wang cracked the *Forbes* "Richest in Tech" list in 2017, at the age of 37, as Asia's youngest tech billionaire, worth some $3.2 billion. The publicity-shy Wang, with his circular glasses, tuft of chin stubble, and golf cap became the world's first drone billionaire. That wealth comes from Wang's ownership of approximately 45 percent of DJI shares and the company's profitability.

Compared to other Chinese tech companies, DJI has raised relatively small amounts of capital but at a much higher valuation—this is generally a good thing for a startup's health. DJI ranks within the top 20 of unicorn-valued companies[3] from its most recent funding of $105 million at a $10 billion valuation. DJI started out with angel investment from a family friend and progressed from there. DJI pulled in around $30 million at a valuation of $1.6 billion in January 2015 from Sequoia Capital China, then $75 million at an $8 billion valuation in May 2015 from Silicon Valley–based Accel Partners, the same firm that backed Facebook and Dropbox. DJI is next aiming to haul in $1 billion at a valuation as high as $15 billion. What follows could be an initial public offering.

Shenzhen Base Makes Sense

DJI is located in Shenzhen for a reason. This former fishing village rose to become the world's factory for Apple iPhone and Nike sneakers and has moved up the ladder to design and development of highly advanced technological products such as drones and other internet-connected devices. DJI's proximity to designers and component suppliers lets it do rapid prototyping to find out what concepts work in practice, scrap those that don't work, and perfect those that do. DJI can design and test its drones within one day and ship them out with little time lost. This gives DJI a competitive advantage in cost of capital, manufacturing, and distribution.

Living up to its full name, Da-Jiang Innovations—a play on the Chinese adage "Great ambition has no boundaries"—has developed a series of innovative drones. There's the mass market $679 Phantom in 2013, the $799 Mavic Pro in 2016 as the first-of-its-kind foldable drone, and the $499 mini Spark in 2017—the first drone that can be controlled simply by hand gestures. With each product success in quick order, DJI solidified its lead, thanks to a 1500-strong research and development team.

DJI stands out from many China-based tech startups for its international footprint. The drone maker has offices in the Netherlands, Australia, Japan, Korea, Germany, and the United States—in Los Angeles. Some 85 percent of its revenues come from international markets, but China is a good base. China's innovative culture and willingness to try new concepts coupled with the country's need for logistics improvements in a vast geography and governmental support make for a fertile market for drone technologies. Global growth in the drone market could be impacted by the US-China trade war, points out Jing Bing Zhang, research director who leads IDC Worldwide Robotics, but he expects a pickup from 2020 onward.[4]

Chinese drone maker DJI has been riding on a fascination with drones. Crowds gather around drone displays at consumer electronic shows and parents give hobby drones to their kids as gifts.

DJI is smartly going after the high-potential commercial drone market, which accounts for about 60 percent of total worldwide sales, used for such purposes as crop spraying, power line inspections, and mapping. DJI's corporate customers in the United States include American Airlines, the Union Pacific Railroad, freight rail network BNSF, as well as West Coast flight software maker AutoModality.

Throughout its short history, DJI has had to beat back interlopers. One competitor that DJI fought was American action camera maker GoPro, anchored in Silicon Valley. GoPro launched its drone Karma in late 2016 but when Karma failed to gain traction, quit the market in early 2018. GoPro was too late with a market entry, and more than that, its drone didn't measure up to DJI's lighter, smaller, and cheaper drone with a longer flight time. DJI also chased away another key rival, 3D Robotics in Berkeley, California. The cofounders were the former *Wired* magazine editor Chris Anderson and Colin Guinn, the former North American head of DJI who was embroiled in a dispute with founder Wang over who was most responsible for the Phantom's success. Wang bought out Guinn's stake in DJI in 2013, shifted all operations to China in 2013 and reached $130 million in revenues that year, turning a profit in 2014. In 2016, the rival 3D Robotics quit making drones and transitioned to software. Meanwhile, sales of French drone maker Parrot have nosedived, shares have dipped, and its majority owner has launched a takeover of the business.

DJI is fending off yet more drones, including California-based Impossible Aerospace, founded by Tesla and SpaceX veterans. The Tesla version came out of stealth mode in late 2018 and has raised $11 million from two Silicon Valley investment firms, Bessemer Venture Partners and Eclipse Ventures. Impossible Aerospace is aiming to upend the status quo with an electric model that can fly for two hours and has been sold to police departments, firefighters, and search and rescue teams.

Can China innovation beat America's champion Tesla in flight? So far, the West has proven to be no match for Chinese drone maker DJI and its high altitude.

AFTERWORD

What China's great leap forward in the global tech economy
means for the United States and its future leadership.

PREDICTING CHINA'S TECH FUTURE

Superpowers United States and China are competing for global dominance of world-changing technologies. It's a pivotal moment. No country stays in power forever.

In the ebb and flow of history, economic powers shift from one country to the next. I believe we are now at this juncture with the United States and China.

Game-changing technologies are being invented in China at a rapid clip, and they're going global. The future of tomorrow is being driven by new economy breakthroughs, largely in high tech, which is transforming our world.

China has the advantage to lead because of its large online markets and a young, tech-savvy population eager to experiment with new devices. There's no legacy of outdated personal computers or dial-up internet. It's straight mobile all the time, and 5G superspeedy connections are on their way.

Chinese tech entrepreneurs are already charged up enough— crazy work schedules, fire in their bellies, ambition with no boundaries, passion with no end.

They make Silicon Valley entrepreneurs look sleepy.

China's world-changing tech sector benefits from government

support pushing key technological sectors to be first in the world. China also gets the benefit of Silicon Valley venture money funding their new ideas and scaling them to world-class tech leaders that go public by the dozens on Nasdaq and the New York Stock Exchange. Meanwhile, many American internet companies remain blocked in China or can't win against dominant Chinese competitors. You may dislike China's methods in subsidizing its technology players, acquiring cutting-edge US startups, and learning the secrets of Silicon Valley, but you can't dispute how far and fast China's tech sector has advanced. It is no wonder that the United States is on the defensive with China and that tensions are building over a trade and tech war between these two superpower leaders.

Venture capital and tech entrepreneurship has existed in its own world of cross-border investments and information exchanges between these two world powers. As China continues to rise with game-changing technologies, frictions over leadership and proprietary technologies could break apart the trans-Pacific collaboration and funding, flow and the synergies that have contributed to advances in artificial intelligence, autonomous driving, electric cars, robotics, and communications. China still lacks some fundamental technologies but its "Made in China 2025" policy shows a undeniable determination to make progress in deep tech by mandating it with broad policy initiatives. The Middle Kingdom could very well become self-sufficient, particularly given the ingenuity and entrepreneurial instincts of the Chinese people—and how fast their digitally savvy Millennials adapt to new tech.

Yes, there are many gaps and social ills, but China is making progress fast. I never could have imagined even just 10 years ago how advanced it would become, how giant Baidu, Alibaba, and Tencent would grow across broad sweeps of the economy. Now a new group of technologically advanced Chinese companies led by serial entrepreneurs are coming up, with their own breakthroughs. Imagine what the next decade could bring.

Power and money are being ingrained into China's culture, with a new-found confidence. It's a pivotal moment, and US policy makers and Silicon Valley tech leaders are well aware that the currents are changing. China's rapid rise as a tech superpower is challenging America's status as the world's technology leader and could lead to a shift in global economic dominance.

Acknowledgments

A lot has changed since my first book, *Silicon Dragon-How China is Winning the Tech Race,* was published in 2008. Back then, the story of China's rise in technology was not readily acknowledged. I was a lone journalist chasing the story.

But now, the timing is right for *"Tech Titans: How China's Tech Sector is Changing the World by Working Harder, Innovating Faster & Going Global."* This topic has gone mainstream. The US-China tech war makes the headlines every day. It's an issue that is debated often now.

Ten years ago, I never could have imagined that the Silicon Dragon theme would become so strong, that a Silicon Dragon ecosystem would grow so quickly that challenges Silicon Valley. I never could have dreamt of leading a Silicon Dragon media platform of news, events and thought leadership. Without the support of the Silicon Dragon community for my entrepreneurial activities, *Tech Titans of China* may not have been possible.

Thank you, Hachette's Nicholas Brealey, for publishing this book globally, and to my book editor Alison Hankey for her vote of confidence and steadfastness in seeing this through. Thank you to my agent and former book editor Leah Spiro for providing insightful guidance on framing this controversial and timely topic. Thank you to the production, proofing and sales teams at Hachette Book Group for keeping this book on track and getting it into readers' hands. I'd also like to thank my editors at Forbes.com and CNBC.com for publishing my regular columns and articles over the years, and to Pulitzer-owned *International Business*, which first sent me to Hong Kong on assignment years ago.

I wish to thank all the many supporters of Silicon Dragon since its start in 2010 in Silicon Valley, New York, and Beijing, and then expansion to multiple innovation hubs. These are the venture capitalists, entrepreneurs, deal makers, professional service firms, corporate leaders, founders, and investment groups, which have been with me on the journey that has led to this book.

You may recognize some of these supporters and sponsors in the pages of this book. I'd like to give a 'shout out' to those who have contributed time, resources and efforts to speaking at our programs, joining our Circle membership, and doing interviews with me many times. I will name a few venture capital firms here that deserve special recognition: Qiming Venture Partners, GGV Capital, Sequoia Capital China, DCM Ventures, Lightspeed Partners China, RRE Ventures, NEA, ZhenFund, Redpoint China Ventures, Long Hill Capital, China Creation Ventures, China Growth Capital, and more.

There are so many more supporters, and I urge you to check out www.silicondragonventures.com to see who else has contributed.

Doing this kind of reporting, writing, programming, and producing requires frequent travels and quick adjustment to widely varying time zones. But I really don't live out of a suitcase or on a plane like some people think. I never could have finished this book without the home base support of my husband, John, my family and close friends, and the tranquility of my hideaway in the San Francisco Bay Area. I've gotten over jet lag a long time ago—a good thing. I am not as lonely on the China venture and entrepreneurial trail as I was 10 years ago. It's good to have company. I hope you enjoy the book!

Notes

Introduction

1. Rebecca A. Fannin, *Silicon Dragon* (New York: McGraw-Hill, 2008); Rebecca A. Fannin, *Startup Asia* (New York: John Wiley & Sons, 2011).
2. Max J. Zenglein and Anna Holzmann, "Evolving Made in China 2025," presented at Asia Society conference, Stanford University, January 15, 2019; asiasociety.org/sites/default/files/2019-01/MERICS%20Evolving%20Made%20 in%20China%202025%20Preview.pdf.
3. Boston Consulting Group, "Decoding the Chinese Internet"; bcg.com/d/press/ 28september2017-decoding-chinese-internet-172187.
4. Mobile Apps Struggle To Retain Most Users In China, eMarketer, December 21, 2015 https://www.emarketer.com/Article/Mobile-Apps-Struggle-Retain-Most -Users-China/1013366
5. Max J. Zenglein and Anna Holzmann, "Evolving Made in China 2025," presented at Asia Society, Stanford University, January 15, 2019; asiasociety. org/sites/default/files/2019-01/MERICS%20Evolving%20Made%20in%20 China%202025%20Preview.pdf.
6. US-China Economic and Security Review Commission, "The 13th Five-Year Plan," February 14, 2017; uscc.gov/Research/13th-five-year-plan. 2018 Report to Congress of the US-China Economic and Security Review Commission, November 14, 2018; uscc.gov/Annual_Reports/2018-annual-report.
7. US-China Economic and Security Review Commission, "The 13th Five-Year Plan," February 14, 2017; uscc.gov/Research/13th-five-year-plan.
8. People's Republic of China, "13th Five-Year Plan on National Economic and Social Development," March 17, 2016. Translation; *gov.cn/xinwen/2016-03/17/ content_5054992.htm.*
9. "China's New $15 Billion Tech Fund Emulates SoftBank's Vision Fund, *The Economist,* July 5, 2018; economist.com/business/2018/07/05/chinas-new-15bn -tech-fund-emulates-softbanks-vision-fund.

10. Defense Innovation Unit Experimental (DiuX), China's Technology Transfer Strategy, January 2018; admin.govexec.com/media/diux_chinatechnology transferstudy_jan_2018_(1).pdf.

11. "Global Venture Capital Trends, Analysis of 2010–2018 Data, London-Based Alternative Assets," Preqin. Accessed January 10, 2019.

12. Jason D. Rowley, "Q4 2018 Closes Out a Record Year for the Global VC Market, *Crunchbase,* January 7, 2019; news.crunchbase.com/news/q4-2018 -closes-out-a-record-year-for-the-global-vc-market/.

13. The Global Unicorn Club, *CB Insights,* customized research; cbinsights .com/research-unicorn-companies.

14. S&P Global Market Intelligence, customized research; spglobal.com/market intelligence/en/client-segments/investment-banking-private-equity.

15. IPO Stats, Renaissance Capital, customized research; renaissancecapital.com/ IPO-Center/Stats. China IPOs in the US accounted for 16 percent of 190 public offerings overall stateside that raised $47 billion in 2018. In 2017, 16 Chinese companies went public and raised $3.3 billion.

16. Data from S&P Global Market Intelligence, customized research; spglobal.com/ marketintelligence/en/client-segments/investment-banking-private-equity. China dealmakers invested $51.4 billion in 586 deals in the United States in 2018.

17. American Association for the Advancement of Science (AAAS), Historical Trends in Federal R&D, April 2018; aaas.org/programs/r-d-budget-and-policy/ historical-trends-federal-rd; PRC National Bureau of Statistics, Chinese Government R&D spending; stats.gov.cn/tjsj/tjgb/rdpcgb/qgkjjftrtjgb/.

18. National Science Board, Science & Engineering Indicators 2018. The NSB had predicted that by 2019, China would surpass the United States, if current annual growth rates continue—China at 18 percent to $409 billion, the United States at 4 percent; nsf.gov/nsb/sei/one-pagers/China-2018.pdf; nsf.gov/ statistics/2018/nsb20181/report/sections/overview/r-d-expenditures-and-r-d -intensity.

19. Main Science and Technology Indicators, Organisation for Economic Cooperation and Development (OECD). China ascended to second place globally in gross domestic expenditures on research and development compared to relatively steady levels for world leader United States; oecd.org/sti/inno/ DataBrief_MSTI_2017.pdf; Randy Showstack, "China Catching Up to US in R&D," National Science Board report. EOS, January 24, 2018; eos.org/ articles/china-catching-up-to-united-states-in-research-and-development.

20. The world's top 20 companies by R&D Investment in 2018. The 2018 EU Industrial R&D Investment Scoreboard. December 17, 2018; ec.europa.eu/ info/news/2018-industrial-rd-scoreboard-eu-companies-increase-research

-investment-amidst-global-technological-race-2018-dec-17_en.https://ec.europa
.eu/info/news/2018-industrial-rd-scoreboard-eu-companies-increase-research
-investment-amidst-global-technological-race-2018-dec-17_en.

21. "Innovators File Record Number of International Patent Applications, with Asia Leading. World Intellectual Property Organization (WIPO), March 19, 2019. https://www.wipo.int/pressroom/en/articles/2019/article_0004.html.

22. "China Drives International Patent Applications to Record Heights," WIPO, March 21, 2018; wipo.int/pressroom/en/articles/2018/article_0002.html. Innovators File Record Number of International Patent Applications, with Asia Now Leading. WIPO, March 19, 2019. https://www.wipo.int/pressroom/en/articles/2019/article_0004.html.

23. "World Intellectual Property Indicators," WIPO, December 3, 2018; wipo.int/pressroom/en/articles/2018/article_0012.html.

24. World Economic Forum, "The Human Capital Report 2016"; weforum.org/docs/HCR2016_Main_Report.pdf.

25. National Science Board, Science & Engineering Indicators 2018, "The Rise of China in Science and Engineering"; nsf.gov/statistics/2018/nsb20181/report.

26. Ibid.

27. Internet World Stats, March 31, 2019; https://www.internetworldstats.com/top20.htm Top 50 Countries/Markets by Smartphone Users and Penetration, 2018 figures from Global Mobile Market Report, Newzoo, Accessed April 18, 2019; https://newzoo.com/insights/rankings/top-50-countries-by-smartphone-penetration-and-users/.

28. Internet World Stats, March 31, 2019; https://www.internetworldstats.com/top20.htm.

29. Robert Castellano, "US Restricts Exports of Some Chip Production Equipment to China," *Seeking Alpha,* November 6, 2018; seekingalpha.com/article/4218617-u-s-restricts-exports-chip-production-equipment-china-impact-memory-equipment-suppliers.

30. Merit Janow and Rebecca A. Fannin, "China: Trade, Tech and Tolerance," Techonomy NYC Conference, May 14, 2018; techonomy.com/conf/nyc18/media-marketing-trade/china-trade-tech-tolerance/.

31. Deloitte, "5G: The Chance to Lead for a Decade," 2018; deloitte.com/content/dam/Deloitte/us/Documents/technology-media-telecommunications/us-tmt-5g-deployment-imperative.pdf.

32. Nic Fildes and Louise Lucas, "Huawei Spat Comes as China Races Ahead in 5G," *Financial Times,* December 12, 2018; can be accessed through subscription at: ft.com/content/0531458a-fd6c-11e8-ac00-57a2a826423e.

33. Mike Cherney and Dan Strumpf, "Taking Cue from the US, Australia Bans Huawei from 5G Network," *Wall Street Journal,* August 23, 2018; can be accessed

through subscription at: wsj.com/articles/australia-bans-chinas-huawei-from
-5g-network-rollout-1534992631.

34. "2017 Venture Capital Deals," Preqin, January 4, 2018. Didi was backed by China's Bank of Communications and China Merchants Bank; docs.preqin.com/press/VC-Deals-2017.pdf.

35. Data from S&P Global Market Intelligence, customized research, accessed January 14, 2019; spglobal.com/marketintelligence/en/client-segments/investment-banking-private-equity.

36. Lorand Laskai, "Why Does Everyone Hate Made in China 2025?" Council on Foreign Relations, March 28, 2018; cfr.org/blog/why-does-everyone-hate-made-china-2025.

Chapter One

1. Fang Ruan, et al., "Year 2035, 400 Million Job Opportunities in the Digital Age," Boston Consulting Group, March 2017; mage-src.bcg.com/Images/BCG_Year-2035_400-Million-Job-Opportunities-Digital%20Age_ENG_Mar2017_tcm52-153963.pdf.

2. "The Billionaires 2019," Forbes, March 5, 2019; https://www.forbes.com/billionaires/#5db48eb7251c3.

3. Market Cap Ranking 2018, Capital IQ, customized research, accessed January 16, 2019; spglobal.com/marketintelligence/en/client-segments/investment-banking-private-equity.

4. "Digital China—Powering the Economy to Global Competitiveness," McKinsey Global Institute, December 2017; mckinsey.com/featured-insights/china/digital-china-powering-the-economy-to-global-competitiveness.

5. Arjun Kharpal, "Alipay's Parent Company Says Tech Services—Not Payments—Will Be Its Main Business in the Future," CNBC, December 4, 2018; cnbc.com/2018/11/14/alipay-parent-ant-financial-says-services-to-surpass-payments-business.html.

6. Harrison Jacobs, "One Photo Shows China Is Already in a Cashless Future," Business Insider, May 29, 2018; businessinsider.com/alipay-wechat-pay-china-mobile-payments-street-vendors-musicians-2018-5.

7. Global OEM Pay Users, Juniper Research, June 2018; juniperresearch.com/press/press-releases/apple-pay-accounts-for-1-in-2-oem-pay-users.

8. Shan Li and Maria Armental, "China's Baidu Credits Artificial Intelligence for Robust Ad Sales," Wall Street Journal, August 1, 2018; can be accessed through subscription at: wsj.com/articles/baidu-reports-strong-quarterly-results-1533085721.

9. Elizabeth Weise, "Amazon Rakes in Estimated $3.5 Billion for Prime Day," USA

Today, July 17, 2018; usatoday.com/story/tech/talkingtech/2018/07/17/amazon-estimated-sell-2-4-billion-since-start-prime-day/792466002/.

10. Gabriel Wildau and Yizhen Jia, *Financial Times*, January 28, 2019; can be accessed through subscription at: ft.com/content/35bbbef6-20a8-11e9-b126-46fc3ad87c65.

11. Ibid.

12. "Where Alibaba and Tencent Got Their Names," WBUR, March 25, 2014; wbur.org/onpoint/2014/03/25/where-alibaba-and-tencent-got-their-names.

13. Dean Takahashi, "The Dean Beat: Tencent Leads China's Domination of the Global Games Business," *Venture Beat*, April 20, 2018; venturebeat.com/2018/04/20/the-deanbeat-tencent-leads-chinas-domination-of-the-global-games-business/.

14. Tencent Music Entertainment prospectus, October 2018; sec.gov/Archives/edgar/data/1744676/000119312518290581/d624633df1.htm.

15. "WeChat 'Mini-Program' Initiative Hits 1 Million Apps," *EJI Insight,* November 8, 2018; ejinsight.com/20181108-wechat-mini-program-initiative-hits-one-million-apps/.

16. Rebecca A. Fannin, "China Releases a BAT," *Techonomy*, May 23, 2018; techonomy.com/2018/05/china-releases-tech-dragon-bat/.

17. Ann Lee, *Will China's Economy Collapse?* (New York: John Wiley & Sons, 2017).

Chapter Two

1. Mike Moritz, "China Is Winning the Global Tech Race," *Financial Times*, June 17, 2018; can be accessed through subscription at: ft.com/content/3530f178-6e50-11e8-8863-a9bb262c5f53. "Between 2015 and 2017, the five biggest US tech groups (especially Apple and Microsoft) spent $228 billion on stock buybacks and dividends. During the same period, the top five Chinese tech companies spent just $10.7 billion and ploughed the rest of their excess cash into investments that broaden their footprint and influence."

2. Data from S&P Global Market Intelligence, customized research, accessed January 14, 2019; spglobal.com/marketintelligence/en/client-segments/investment-banking-private-equity.

3. Ibid.

4. Ibid.

5. Thilo Hanemann, "Arrested Development: China FDI in the US in 1H 2018," Rhodium Group, June 19, 2018; rhg.com/research/arrested-development-chinese-fdi-in-the-us-in-1h-2018/.

6. Thilo Hanemann, Cassie Gao, Adam Lysenko, "Net Negative: Chinese Investment in the US in 2018," Rhodium Group, January 13, 2019; rhg.com/research/chinese-investment-in-the-us-2018-recap/.

7. Thilo Hanemann, Daniel Rosen, Cassie Gao, "Two-Way Street, 2018 Update, US-China Direct Investment Trends," Rhodium Group, National Committee on US China Relations; rhg.com/research/two-way-street-2018-update-us -china-direct-investment-trends/.

8. Rebecca A. Fannin, "China to US Tech Investment Plunges 79%," Forbes, January 21, 2019; forbes.com/sites/rebeccafannin/2019/01/21/china-to-us-tech-investment-plunges-79-to-lowest-level-in-7-years-amid-dc-crackdown/#79b 9ce371964.

9. CFIUS, "China Deals That Can Still Be Done," Pillsbury Winthrop Shaw Pittman, October 26, 2018; jdsupra.com/legalnews/cfius-china-deals-that-can-still -be-done-16618/.

10. Hans Tung and Zara Zhang, "Why Chinese Entrepreneurs are Targeting Emerging Markets Across the World," 996 (podcast), episode 28; 996.ggvc .com/2019/01/16/episode-28-chuhai-why-chinese-entrepreneurs-are-targeting -emerging-markets-across-the-world/.

11. China Outbound Tech Investment, customized data, 2011–2017, Standard & Poor's research division Capital IQ, S&P Global Market Intelligence, customized data, Accessed January 13, 2019. The proportion for Asia is growing while the US share is shrinking—to about 8 percent in 2018 from almost half (43 percent) in the peak year of 2016; spglobal.com/marketintelligence/ en/client-segments/investment-banking-private-equity.

12. Rebecca A. Fannin, Startup Asia (New York: John Wiley & Sons, 2011).

13. Tim Merel, "Digi-Capital: $5.7 Billion Games Investment in 2018 Double Previous Record," Yahoo! Finance, February 4, 2019. Tencent invested $630 million in Douyu, $474 million in Shanda Games, and $462 million in Huya; finance .yahoo.com/news/digi-capital-5-7-billion-144100171.html.

14. Sherisse Pham, "Tencent Pumps Billions into 300 Companies. Here's What It's Buying," CNN Business, October 4, 2018; cnn.com/2018/10/04/tech/tencent-investment-strategy-explained/index.html.

15. Celia Chen and Iris Deng, "Has Tencent Lost Is Creative Mojo?" South China Morning Post, May 7, 2018; scmp.com/tech/tech-leaders-and-founders/article/ 2144973/has-tencent-lost-its-creative-mojo-essay-sparks.

16. Matthew Brennan, Tencent's Investment Strategies Revealed, China Channel, August 25, 2018; chinachannel.co/tencents-investment-strategies-revealed/.

Chapter Three

1. Jon Russell, "Xiaomi's Mi8 May Be Its Most Brazen iPhone Copycat Yet," Tech-Crunch, May 31, 2018; techcrunch.com/2018/05/31/xiaomis-mi-8-may-be-its -most-brazen-iphone-copycat-yet/.

2. Paul Carsten, "For China's Xiaomi, It's What's Inside That Counts," Reuters, August 16, 2013; reuters.com/article/us-xiaomi-china/for-chinas-xiaomi-its-whats-on-the -inside-that-counts-idUSBRE97F04420130816.

3. IDC statistics, Industry Overview, first quarter 2018, Xiaomi Global Offering, June 25, 2018; hkexnews.hk/listedco/listconews/sehk/2018/0625/ ltn20180625033.pdf.

4. Ronald Keung, "China E-Commerce: The Next Leg of Growth," Goldman Sachs, July 13, 2017; goldmansachs.com/insights/pages/ronald-keung-china-next-leg-of -growth.html.

5. "Billionaires: Lei Jun: #11," Forbes, February 12, 2019; forbes.com/ profile/lei-jun/#6de73f006e64.

6. IDC, "China Smartphone Units Drop by 6% YoY," August 2018; www.idc.com/ getdoc.jsp?containerId=prCHE44199418.

7. Ben Thompson, "Xiaomi's Ambition," Stratechery, January 7, 2015; stratechery .com/2015/xiaomis-ambition/.

8. Xiaomi Lei Jun, "Let Us Witness a Great Milestone Together," KrAsia, July 8, 2018; kr-asia.com/xiaomi-lei-jun-let-us-witness-a-great-milestone-together; "In IPO letter, Xiaomi CEO Explains Innovation at Honest Prices," Bloomberg News, May 3, 2018; bloomberg.com/news/articles/2018-05-03/in-ipo-letter-xiaomi -ceo-explains-innovation-at-honest-prices.

9. ByteDance overview, home page, bytedance.com/#about0.

10. Qian Chen, "The Biggest Trend in Chinese Social Media Is Dying," CNBC, September 18, 2018; cnbc.com/2018/09/19/short-video-apps-like-douyin-tiktok -are-dominating-chinese-screens.html.

11. "Apple Reveals the Most Popular iPhone Apps of 2018," Mashable, December. 4, 2018; mashable.com/article/apple-most-popular-iphone-apps-2018/#Dal8P .VlBGqn.

12. Alex Fang, "China's Richest 2018," Forbes, October 24, 2018; forbes.com/sites/ alexfang/2018/10/24/15-under-40-meet-the-youngest-members-of-chinas-400 -richest/#529ff5277369.

13. Qian Chen, "The Biggest Trend in Chinese Social Media Is Dying, and Another Is Already Taking Its Place," CNBC, September 19, 2018; cnbc. com/2018/09/19/short-video-apps-like-douyin-tiktok-are-dominating-chinese- screens.html.

14. Paul Armstrong and Yue Wang, "The Billion-Dollar Race to Become the Net- flix of China," Forbes, May 7, 2018; forbes.com/sites/ywang/2018/03/07/the -billion-dollar-race-to-become-the-netflix-of-china/.

15. Qian Chen, "The Biggest Trend in Chinese Social Media Is Dying," CNBC, Sep- tember 18, 2018; cnbc.com/2018/09/19/short-video-apps-like-douyin-tiktok -are-dominating-chinese-screens.html.

16. Michael K. Spencer, "Who Is Yiming Zhang? The Founder of ByteDance Is Virtually Unknown Outside of China," *Medium*, August 23, 2018; medium .com/futuresin/who-is-yiming-zhang-595a52c8ffb1.

17. Josh Horwitz, "China's Next Generation Giants Are Getting Big by Serving the Country's Poor," *Quartz,* July 19, 2018; qz.com/1331677/tech-startups -bytedance-kuaishou-getting-big-serving-chinas-poor/.

18. Connie Chan, "When AI Is the Product: The Rise of AI-Based Consumer Apps," Andreessen Horowitz, December 3, 2018; a16z.com/2018/12/03/ when-ai-is-the-product-the-rise-of-ai-based-consumer-apps/.

19. Anu Hariharan, "The Hidden Forces Behind Toutiao," YCombinator, October 12, 2017; blog.ycombinator.com/the-hidden-forces-behind-toutiao-chinas-content -king/.

20. Meituan Dianping Global Offering, prospectus, September 7, 2018; http:// www3.hkexnews.hk/listedco/listconews/sehk/2018/0907/ltn20180907011.pdf. Accessed December 1, 2018.

21. Louise Lucas, "Chinese Food Delivery App Meituan Posts Widening Losses," *Financial Times*, November 22, 2018; can be accessed through subscription at: ft.com/content/90e2a8ac-ee3b-11e8-89c8-d36339d835c0.

22. Introduction of Meituan Dianping Group, PowerPoint, October 2018. Meituan Dianping Results for the Year Ended 2018, March 11, 2019. http://meituan. todayir.com/attachment/201903111815020000342 2038_en.pdf. Accessed December 1, 2018.

23. "McKinsey Insights China: Meet the 2020 Chinese Consumers, March 2012," mckinsey.com/featured-insights/asia-pacific/meet-the-chinese-consumer-of-2020.

24. Lily Varon, et al., "eCommerce in China: Trends and Outlook for the Largest eCommerce Market in the World, Forrester, August 10, 2018; forrester .com/report/eCommerce+In+China+Trends+And+Outlook+For+The+Largest +eCommerce+Market+In+The+World/-/E-RES143994#.

25. Meituan IPO prospectus, iResearch Global, September 7, 2018; hkexnews.hk/ listedco/listconews/sehk/2018/0907/ltn20180907011.pdf.

26. Eleanor Creagh, "Equities: A Closer Look at Meituan Dianping," Saxo Capital Markets, September 8, 2018; home.saxo/insights/content-hub/articles/2018/09/07/ a-closer-look-at-meituan-dianping.

27. iResearch Global, "Report of China's Food Delivery Apps in First Half 2018," August 15, 2018; iresearchchina.com/content/details8_46801.html.

28. "China Rich List 2018: Wang Xing, #37," *Forbes*, February 18, 2019; forbes.com/ profile/wang-xing/#79963f291686.

29. Lucinda Shen, "Meituan Shares Are Down 27% in 2018. Here's Why Sequoia Has Not Sold a Single Share," *Fortune,* November 29, 2018; fortune .com/2018/11/29/india-sequoia-capital-meituan-dianping/.

Chapter Four

1. "Starbucks and Alibaba Announce Partnership to Transform the Coffee Experience in China," August 2, 2018; stories.starbucks.com/press/2018/starbucks -and-alibaba-announce-partnership-to-transform-coffee-experience/.

2. Benjamin Romano, "Starbucks Trying to Value the Dignity of Work, Schultz Tells Shareholders," *Seattle Times*, March 21, 2018; seattletimes.com/business/ starbucks/starbucks-trying-to-value-the-dignity-of-work-schultz-tells -shareholders/.

3. "Why Bill Ackman and Coca-Cola Are Betting Big on Coffee in China," *Bloomberg News*, October 12, 2018; bloomberg.com/news/articles/2018-10-10/ why-bill-ackman-and-coca-cola-are-betting-big-on-coffee-in-china.
China Coffee Market, October 2018. Mintel. Accessed December 1, 2018; store. mintel.com/china-coffee-market-report. Matthew Berry, "China Hot Drinks Industry," *Euromonitor*, January 17, 2018; blog.euromonitor.com/china-hot -drinks-industry/. "Coffee in China, Country Report," *Euromonitor*, March 13, 2019; euromonitor.com/coffee-in-china/report.

4. International Coffee Organization and the U.S. Department of Agriculture, Coffee consumption grew by 16 percent annually over the past decade compared with 2 percent growth on average worldwide; ico.org/trade_statistics .asp?section=Statistics.

5. Pei Li and Adam Jourdan, "Coffee Startup Luckin Plans to Overtake Starbucks in China," Reuters, January 3, 2019; reuters.com/article/us-china-coffee -luckin/coffee-startup-luckin-plans-to-overtake-starbucks-in-china-this-year -idUSKCN1OX0BY.

6. Mike Isaac, "Facebook Said to Create Censorship Tool to Get into China," *New York Times*, November 22, 2016; nytimes.com/2016/11/22/technology/facebook-censorship-tool-china.html.

7. Jeff Weiner, "LinkedIn in China: Connecting the World's Professionals," Linkedin, February 24, 2014; linkedin.com/pulse/20140224235450-22330283 -linkedin-in-china-connecting-the-world-s-professionals/.

8. Dominic Penaloza, "A New Tool for Living a Life Well Lived," *Medium*, February 11, 2016; medium.com/@domthecalm/a-new-tool-for-living-a-life-well-lived -the-40-percent-rule-3b8f7ec47826. Dominic Penaloza, "What Happened to Ushi.com…" *Quora*, October 6, 2016; quora.com/What-happened-to-Ushi-com-Ushi-cn-Does-it-still-exist.

9. "China's Co-working King Ucommune Leverages Smart Tech to Compete with WeWork," Silicon Dragon on YouTube, November 18, 2018; youtube.com/ watch?v=yJj1DHZQOOo.

10. Chenyu Zheng, "Six Jaw-Dropping Airbnb Homes to Experience China,"

Medium, September 17, 2016; medium.com/@chenyuz/6-jaw-dropping-airbnb -homes-to-experience-the-authentic-and-historic-china-ad4f7e87aa0e.

11. Xinhua, "How Tourism Is Becoming a New Driving Force in China's Growth," *China Daily*, March 5, 2018. Inbound and outbound travelers to China numbered 123 million, and grew nearly 16 percent in 2017 to a $720 billion market; http://www.chinadaily.com.cn/a/201803/05/WS5a9d08eda3106e7dcc13faad. html.

12. Deanna Ting, "Airbnb China Names New President," *Skift,* July 10, 2018; skift .com/2018/07/10/airbnb-china-names-new-president/.

13. Hans Tang and Zara Zheng, "Nathan Blecharczyk on Lessons from Airbnb's China," *996* (podcast), episode 10, April 11, 2018; 996.ggvc.com/2018/04/11/ episode-10-nathan-blecharczyk-on-lessons-from-airbnbs-china-expansion/.

Chapter Five

1. Silicon Dragon 2018, China VC panel, November 15, 2018; silicondragonventures.com/videos/silicon-dragon-vc-awards-2018-china-vc-panel/.

2. "China Rich List 2019: Neil Shen, #1," *Forbes*; forbes.com/profile/neil-shen/ #9d4a61d21ff0.

3. Alex Konrad, "The Best Dealmakers in High-Tech Venture Capital in 2019," *Forbes,* April 2, 2019; *www.forbes.com/midas/.*

4. Dan Frommer, "What Happened to TechCrunch's Sequoia China Bribery Allegations?" *Business Insider,* May 18, 2009; businessinsider.com/did-sequoia-threaten -to-sue-techcrunch-2009-5.

5. Yuliya Chernova, "Sequoia Capital Goes on Fund-Raising Spree," *Wall Street Journal*, March 5, 2018; wsj.com/articles/sequoia-capital-goes-on-fundraising -spree-1520253046.

6. Venture capital data, customized research for venture capital investments in China, United States, global markets, and venture funding figures, Preqin. Accessed January 10, 2019.

7. Pramugdha Mamgain, "China Pips US in Attracting Highest VC Funding at $56 billion in H1: Study," *Dealstreet Asia*, December 5, 2018; dealstreetasia.com/ stories/china-pips-us-venture-capital-112507/. In the first half of 2018, China venture investments were $56 billion, topping the United States at $42 billion.

8. China and US venture benchmarking of internal rate of returns for funds, prepared confidentially for Silicon Dragon by a global fund-of-funds investor in private equity and venture capital funds. An analysis of China and US venture fund performance reveals that seven prominent China venture funds earned an average 21.4 percent return, higher than 19.3 percent for 141 US funds but not at the level of 34 percent returns for the top 25 US venture funds.

9. China VC funds outperform United States and Europe, *eFront,* February 20, 2019; efront.com/research-papers/looking-at-chinese-venture-capital-fund-distributions/. A detailed analysis of venture capital funds globally showed that Chinese funds generated 1.79 times returns, slightly higher than US and European funds.

10. "From Alibaba to Zynga: 40 of the Best VC Bets of All Time," *CB Insights,* January 4, 2019; cbinsights.com/research/best-venture-capital-investments/.

11. Jason D. Rowley, "Q4 2018 Closes Out a Record Year for the Global VC Market," *Crunchbase,* January 7, 2019; crunchbase.com/news/q4-2018-closes-out-a-record-year-for-the-global-vc-market/.

12. "The Global Unicorn Club," *CB Insights,* customized research; cbinsights.com/research-unicorn-companies.

13. "China Asserts Itself in the 2018 US IPO market," Renaissance Capital, October 4, 2018; renaissancecapital.com/IPO-Center/News/60165/China-asserts-itself-in-the-2018-US-IPO-market.

14. Research prepared for *Tech Titans of China* by Dealogic, January 11, 2019.

15. "2018 IPO Market Annual Review," Renaissance Capital, December 17, 2018; zy226-894a70.pages.infusionsoft.net/. Plus, Renaissance Capital statistics prepared for *Tech Titans of China*, January 11, 2019.

16. The Dominance of RMB-denominated Funds in China-Focused Private Equity Fund Raising, Preqin, December 2017; preqin.com/insights/blogs/the-dominance-of-rmb-denominated-funds-in-china-focused-private-equity-fundraising/20703. Funds in renminbi, or RMB, increased to $8.4 billion in 2017, from $5.3 billion in 2009.

17. "China's Venture Capital and Private Equity Market Statistics & Analysis," China Venture Institute, accessed February 18, 2019. Of 280 China venture funds raised in 2018, 18 were US dollar funds, with an average size of $386 million compared with $66 million for RMB funds.

18. "China tops the world in incubators," *XinhuaNet,* September 2017; xinhuanet.com//english/2017-09/19/c_136620977.htm.

19. Thilo Hanemann, Cassie Gao, and Adam Lysenko, "Net Negative: Chinese Investment in the US in 2018," Rhodium Group, January 13, 2019; rhg.com/research/chinese-investment-in-the-us-2018-recap/.

20. China and US venture capital deals, customized research for Silicon Dragon by Preqin, a London-based alternative assets source. Accessed January 16, 2019.

21. "Defense Innovation Unit Experimental (DiuX), China's Technology Transfer Strategy," *CB Insights,* January 2018; admin.govexec.com/media/diux_chinat echnologytransferstudy_jan_2018_(1).pdf. China-based investors chalked up 1,201 venture deals and roughly $31 billion in US tech companies, during a

seven-year period ending in October 2017. Moreover, China represented 8 percent of $372 billion in funding of young US technology groups, hitting a peak of 16 percent or $11.5 billion in 2015.

22. Thilo Hanemann, Cassie Gao, and Adam Lysenko, "Net Negative: Chinese Investment in the US in 2018," Rhodium Group, January 13, 2019; rhg.com/research/chinese-investment-in-the-us-2018-recap/.

23. "Venture Capital Funding Report, 2018: PwC and CB Insights' Q4 2018 Money-Tree; cbinsights.com/research/report/venture-capital-q4-2018/.

24. "The Midas List, Top Tech Investors 2018: Jenny Lee, #19," *Forbes,* April 2018; forbes.com/profile/jenny-lee/?list=midas#731d992f18ac.

Chapter Six

1. Zak Doffman, "Chinese Media Claims NYPD Is Using Beijing-Controlled Facial Recognition—Is It True?" *Forbes,* January 13, 2019; forbes.com/sites/zakdoffman/2019/01/13/chinese-media-claims-nypd-is-using-beijing-controlled-facial-recognition-is-it-true/#129e6ca592a6.

2. "Rise of China's Big Tech in AI: What Baidu, Alibaba, and Tencent Are Working On," *CB Insights,* April 26, 2018; cbinsights.com/research/china-baidu-alibaba-tencent-artificial-intelligence-dominance/.

3. "Sizing the Prize: PwC's Global Artificial Intelligence Study: Exploiting the AI Revolution," PwC; pwc.com/gx/en/issues/data-and-analytics/publications/artificial-intelligence-study.html.

4. Kai-Fu Lee, *AI Superpowers* (New York: Houghton Mifflin Harcourt Publishing Co., 2018); aisuperpowers.com/.

5. PwC/CB Insights, "MoneyTree Report, Q4 2018"; pwc.com/us/en/moneytree-report/moneytree-report-q4-2018.pdf.

6. "China Is Starting to Edge Out the US in AI Investment," *CB Insights,* February 12, 2019; cbinsights.com/research/china-artificial-intelligence-investment-startups-tech/.

7. Leisheng Wang, "The Treasure Voyage of Chinese Artificial Intelligence," *China Entrepreneur*, November 12, 2017.

Chapter Seven

1. "China's Mobility Industry Picks Up Speed," Bain & Co., May 2018; bain.com/insights/chinas-mobility-industry-picks-up-speed/.

2. Ibid.

3. Didi overview for Rebecca A. Fannin, based on Didi headquarters visit in Beijing, November 2, 2018.

4. "China's Mobility Industry Picks Up Speed," Bain & Co., May 2018; bain.com/insights/chinas-mobility-industry-picks-up-speed/.

5. Jake Spring, Tatiana Bautzer, and Gram Slattery, "China's Didi Chuxing Buys Control of Brazil's 99 Ride-Hailing App," Reuters, January 3, 2018; reuters.com/article/us-99-m-a-didi/chinas-didi-chuxing-buys-control-of-brazils-99-ride-hailing-app-idUSKBN1ES0SJ.

6. Lulu Yilun Chen, "Beleaguered Didi," *Bloomberg,* September 7, 2018; bloomberg.com/news/articles/2018-09-07/beleaguered-didi-is-said-to-lose-585-million-in-just-six-months. Didi lost some $300 million to $400 million in 2017 and $584 million in the first half of 2018. Didi doled out about $1.7 billion in subsidies and discounts in the first six months of 2018.

7. "Didi Loses $584 million in First Half 2018," *KrAsia,* September 5, 2018; kr-asia.com/exclusive-didi-loses-584m-in-first-half-2018.

8. "Didi Blames 'Ignorance and Pride' for Carpool Murder," BBC News, August 28, 2018; bbc.com/news/world-asia-china-45337860.

Chapter Eight

1. Ma Rui and Lu Ying-Ying, "Pinduoduo: From Zero to $23 Billion in Three Years," TechBuzz China by Pandaily, August 9, 2018; pandaily.com/ep-17-pinduoduo-from-zero-to-23b-in-three-years/.

2. Thomas Graziani, "Pinduoduo: A Close Look at the Fastest-Growing E-commerce App in China," *WALKTHECHAT,* August 28, 2018; walkthechat.com/pinduoduo-close-look-fastest-growing-app-china/.

3. Securities and Exchange Commission, Form F-1, Pinduoduo prospectus, June 29, 2018; sec.gov/Archives/edgar/data/1737806/000104746918004833/a2235994zf-1.htm. Pinduoduo had net losses of approximately $79.5 million in 2017.

4. Erik Schatzker, "Why Blue Orca's Aandahl is Shorting Pinduoduo," Bloomberg Markets and Finance (video), November 14, 2018; youtube.com/watch?v=wtRJY2C-YcY.

5. "E-commerce in China: Trends and Outlook for the Largest E-commerce Market in the World," Forrester, August 10, 2018; forrester.com/report/eCommerce+In+China+Trends+And+Outlook+For+The+Largest+eCommerce+Market+In+The+World/-/E-RES143994#.

6. "China Rich List, 2018: Colin Huang, #12," *Forbes,* April 23, 2019; https://www.forbes.com/profile/colin-huang/#69044eb6d817.

7. Drew Singer and Meghan Genovese, "Pinduoduo to Raise More than $1 Billion in Alibaba Challenge," *Bloomberg,* February 5, 2019; bloomberg.com/news/articles/2019-02-06/china-s-pdd-to-raise-more-than-1-billion-in-alibaba-challenge.

8. Securities and Exchange Commission, Form F-1, Pinduoduo prospectus, June 29, 2018; sec.gov/Archives/edgar/data/1737806/000104746918004833/a2235994zf-1.htm.

Chapter Nine

1. Michael Dunne, "Driving the Future of US-China Relations: China's Global Automotive Push," Asia Society Northern California, February 27, 2019; asiasociety.org/video/driving-future-us-china-relations-chinas-global-automotive-push.
2. China Passenger Car Association, broker reports. Accessed February 28, 2019.
3. Luca Pizzuto et al., "How China Will Help Fuel the Revolution in Autonomous Vehicles," McKinsey & Co., January 2019; mckinsey.com/industries/automotive-and-assembly/our-insights/how-china-will-help-fuel-the-revolution-in-autonomous-vehicles?reload.
4. Dana Hull and Peter Blumberg, "Tesla Joins Apple in Trade Secret Cases Tied to Xpeng, *Bloomberg*, March 21, 2019; bloomberg.com/news/articles/2019-03-21/tesla-sues-rival-zoox-claiming-ex-workers-stole-trade-secrets.
5. Trefor Moss, "Chinese Annual Car Sales Slip for First Time in Decades," *Wall Street Journal*, January 14, 2019; wsj.com/articles/chinese-annual-car-sales-slip-for-first-time-in-decades-11547465112.

Chapter 10

1. "China's Industrial and Military Robotics Development," Defense Group Inc., report prepared for the US-China Economic and Security Review Commission, October 2016; uscc.gov/sites/default/files/Research/DGI_China%27s%20Industrial%20and%20Military%20Robotics%20Development.pdf.
2. "Worldwide Spending on Robotics and Drones Forecast to Total $115.7 Billion," International Data Corp., December 4, 2018: idc.com/getdoc.jsp?containerId=prUS44505618.
3. The Global Unicorn Club, CB Insights, March 1, 2019; cbinsights.com/research-unicorn-companies.
4. Worldwide Spending on Robotics Systems and Drones Forecast, December 4, 2018; idc.com/getdoc.jsp?containerId=prUS44505618.

Index